INDIA'S ANIMALS

Helping The Sacred & The Suffering

Some of the Books by Michael W. Fox

The Whistling Hunters

Concepts in Ethology Animal and Human Behavior

The Soul of the Wolf

The Wild Canids

Behaviour of Wolves Dogs and Related Canids

Integrative Development of Brain and Behaviour in the Dog

Understanding Your Dog

Healing Touch for Cats

Cat Body, Cat Mind

Killer Foods

Beyond Evolution

Superdog

Agricide

The Boundless Circle

Bringing Life to Ethics: Global Bioethics

One Earth One Mind

Animal Rights / Human Rights

Healing Animals & the Vision of One Health

Animals & Nature First

INDIA'S ANIMALS

Helping The Sacred & The Suffering

Deanna.L.Krantz
Dr.Michael.W.Fox

One Health Vision Press

One Health Vision Press, Minneapolis, MN
An.imprint.of.Mozart.&.Reason.Wolfe.Ltd.Wilmington,DE

ISBN-10: 0-911385-67-3
ISBN-13: 978-0-911385-67-0

Book Design and Graphics: D.Krantz & M.W.Fox
Cover Photo: M. W. Fox
Design and Editing Assistance: RaianGarciaCalusa,.A.E.Wittbecker

Printed in the United States of America

First Edition

1 2 3 4 5 6 7 8 9 10

CONTENTS

Preamble
What We Tried to Do

Perhaps against the better judgment of one of us (Fox), after having experienced the difficulties of conducting field studies of the wildlife in India and subsequently giving keynote addresses to the Indian Veterinary Academy and an international gathering convened by the Animal Welfare Board of India, weighed with the responsibility of the other (Krantz), having witnessed first-hand the plight of the animals in this sub-continent, we resolved to return to India and put our concerns into action: First, by establishing an animal shelter and refuge with full veterinary services in an area in dire need, and second, by addressing the endangered status of wildlife and habitat, as well as the lack of enforcement of animal welfare and environmental protection laws.

It turned out to be true that no good deed goes unpunished and that the road to hell is paved with good intentions. Our involvement in India, as the following reports demonstrate, met opposition from various quarters, notably by those with a vested interest in maintaining the status quo of animal exploitation and purported exemplary animal welfare and protection, and who did not want to see the false world-image of India as a spiritual culture, where all living beings are regarded as sacred, tarnished in any way by the facts we uncovered. At least we earned the dedicated support of local people who realized the connections between their well-being and the health and welfare of the animals wild and domesticated.

We collaborated at times with Vandana Shiva, India's leading opponent of GMOs (genetically modified crops), who has been quick to point to this agricultural biotechnology as a new form of corporate colonialism, and who has witnessed how funds from international animal welfare and conservation organizations are misappropriated and how corruption is an endemic problem. However, as foreigners, we were treading on delicate political soils where distrust of our intentions was palpable. We were not intent on shaming India or becoming political agitators for the poor tribals and "scheduled castes," whose animals were neglected by the government. We were not trying to foment unrest—Or, as we were accused by one Minister of the Environment & Animal Welfare, Maneka Gandhi, "out to get rich on the backs of India's suffering animals." She also falsely stated that we had no real animal refuge or shelter, but simply "a country retreat with no animals," when in fact we had cared.for.over.300.sick.or abused animals!

India, with its riches of cultural and natural biodiversity and its ancient legacy of spiritual teachings, music and the arts, but with a burgeoning population and expanding industrialization, is at a crossroads. We hope that choosing the path of compassion and respect for all life—not simply in word but in action and in effective enforcement of established laws that are so urgently needed for domestic and wild animals, as we document in this report—will be the enlightened choice for this nation state. The Light of India may then shine again and inspire the world.

Chapter 1
In The Beginning: Establishing The Refuge & Veterinary Services

On our way to India in 1994 to give the keynote address to the Indian Veterinary Academy in Hissar in the north of this subcontinent, we first visited the Nilgiris in the southern state of Tamil Nadu, where I (Fox) had studied the dholes or wild dogs in the early 1970s. With her background in animal welfare law enforcement and animal shelter operations (formerly with the ASPCA in New York city), Krantz was appalled at the condition of animals on the streets, seeing masses of cattle tied together on the road being force-marched to slaughter, and witnessing the deplorable condition of others being semi-starved at a shelter being run by some Jains under the banner of the Nilgiri Animal Welfare Society (NAWS). So, she met with the chairman of the board, a Mr. Moolchand, who invited her to take it over and make whatever improvements she thought were needed. This she did, surprising them by coming to live at the shelter at her own expense rather than simply sending them money as most donors and Western animal charities had done for decades. This shelter is in heart of the 260 sq. mile Mudumalai wildlife sanctuary, which is part of the Nilgiri Biosphere Reserve.

Nilgiri is the first of the fourteen biosphere reserves of India, established in September 1986, embraces the sanctuary complex of Wyanad, Nagarhole, Bandipur and Mudumalai, the entire forested hill slopes of Nilambur, the Upper Nilgiri plateau, Silent Valley and the Siruvani hills. The total area of the biosphere reserve is around 5520 Sq. Km. The Biosphere reserve is split into three major zones viz. Core Zone, Manipulation forestry Zone, Tourism Zone and Restoration Zone. The break up for the above four zones are as follows: Core Zone, 1240.3 Sq. Km (22.5%); Manipulation forestry Zone, 3238.7 Sq. Km (58.6%); Tourism Zone, 335.0 Sq. Km. (6.1%); and Restoration Zone, 706.4 Sq. Km. (12.8%).

The reserve encompasses 5,520 Sq. Km in the states of Tamil Nadu (2537.6 Sq. Km), Karnataka (1527.4 Sq. Km) and Kerala (1455.4 Sq. Km). It forms an almost complete ring around the Nilgiri Plateau. The biosphere lies between 10°50'N and 12°16'N latitude and 76°00'E to 77°15'E longitude.

The reserve extends from the tropical and subtropical moist broadleaf forests, and tropical moist forests of the western slopes of the Ghats, to the tropical and subtropical dry broadleaf forests and tropical dry forests on the east slopes. Rainfall ranges from 500mm to 7000mm per year. The reserve encompasses three ecoregions, the South Western Ghats moist deciduous forests, South Western Ghats montane rain forests, and South Deccan Plateau dry deciduous forests.

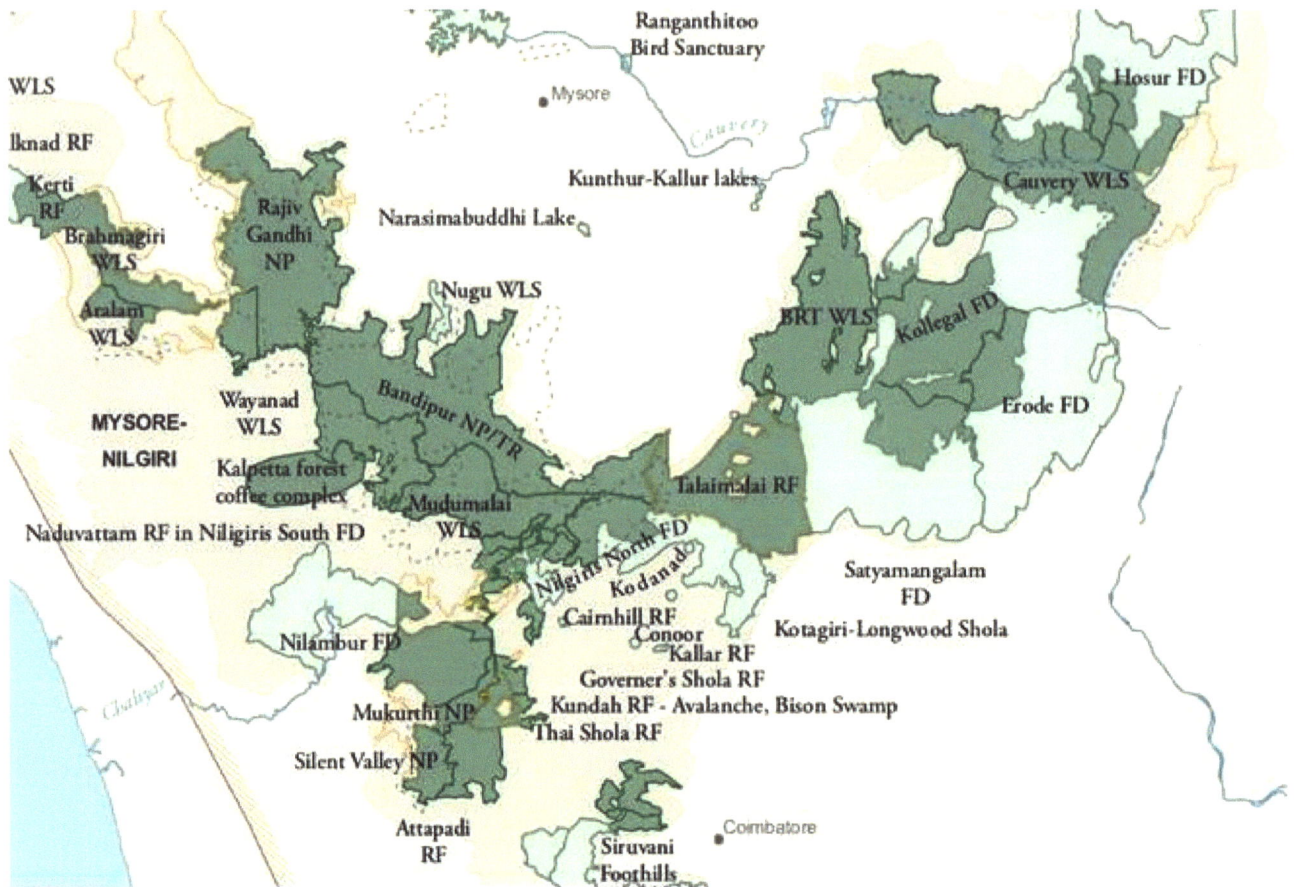

Map of the Nilgiri.

Soon after taking over directing operations and restoring the NAWS shelter and land, Deanna set up the **India Project for Animals & Nature** (IPAN) in 1996 under the auspices of the not-for–profit Global Communications for Conservation Inc., New York (receiving no salary at any time during her 8 years of involvement).

We posted the following in our Mission Statement:

> ***India's animals, the sacred and the suffering, the wild and the domesticated, are in dire need of help***. Help from within this continent of vastly differing cultures and bioregions is extremely limited in part because of the ever pressing demands and problems of a soon-to-be one billion people.

> There is sufficient feed for only **60 percent of India's 200 million cattle**. Veterinary services in many regions lack basic infrastructure and cannot even provide safe and effective vaccines for the half-billion domestic animals in India's 600,000 villages and crowded cities and slums.

> Located in this relatively wild and remote region in the Nilgiri or Blue Mountains, which is one of the oldest mountain ranges in the world, IPAN is a ray of hope for thousands of animals, especially since **most of their owners cannot afford or secure adequate veterinary services**.

In a region with one of the largest domestic animal populations in India, IPAN's program to treat and prevent diseases in these animals directly protects wildlife in the surrounding jungle from the devastating consequences of **contagious diseases**, such as rabies, foot and mouth disease, and distemper.

Here are some of the postings from our Newsletters:

Helping India's Animals: Battles and Victories

As the Fates would have it, I have lost my wife, Deanna Krantz, to India. She has been working there most of the time since 1996 under conditions that even by Indian standards are considered at the lower end of the socio-economic spectrum. She has water which is often "brown" during Monsoons or the dry season, but no shower; a roof but rats and rat poop fall on her bed at night; and on one occasion even a cobra fell from the rafters. She has been victim of people who saw India Project for Animals and Nature (IPAN), which she directs, as a threat.

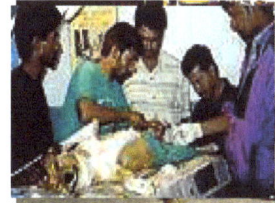
IPAN veterinarian demonstrating spay procedure at community animal hospital.

Street dogs are shown searching for food in the garbage-strewn street.

When you are engaged in corrupt activities like exploiting tribal peoples, forest resources, and animals wild and domesticated, a project like IPAN coming into your fiefdom is unsettling. So you bribe the police, which is common practice, to harass my wife, and you file false charges against her. When she continues her work, providing emergency animal care, humane education, free dog spay-neuter program, and veterinary services in a region of south India where there are no adequate veterinary services for some 8,000 cattle, 4,000 sheep and goats, and 2,000 dogs, you launch a disinformation campaign amongst the villages and tribal settlements. Few people anyway can believe that anyone could be so altruistic as she and not have some ulterior, pecuniary motive.

But Deanna continues undaunted and by sheer dedication, winning the trust of indigenous peoples who see what miracles appropriate veterinary care can accomplish. Even during episodes of sickness, despair, and sheer exhaustion, she continued the work for the animals. An outbreak of rabies in two villages; an epidemic of foot and mouth disease that left calves and cows crippled with their feet infested with maggots; scores of malnourished dogs and pups with mange and rickets, and calves with worms, diarrhea and pneumonia. In addition to running a 24-hour treatment program, often finishing up in some remote tribal settlement at 2:00 a.m. for her Indian veterinarian to do a Caesarian on a cow, she kept up with the rabies and distemper vaccination schedules from village to village. The strength of local support for IPAN is attested by the elected village (Panchayet) government authorities giving her a building to serve as a community animal hospital and for her dog spay-neuter program. Field assistant Nigel Otter, has given over his small plot of land called Hill View Farm, which after some basic renovation, now serves as IPAN's staff quarters and as a refuge for many dogs, some cattle, and over sixty donkeys. These animals came from the defunct Nilgiri Animal Welfare Society's (NAWS) "animal sanctuary" across the river from the farm animal refuge.

The NAWS animal sanctuary used to serve the community with stables for abandoned livestock and a free dispensary. But after the death of its founder, Dorothy Dean, in 1974, the Society ceased to function and the Sanctuary fell into disrepair, becoming a place of animal neglect, cruelty and suffering, along with a variety of unethical and illegal activities. Deanna's attempts to restore this sanctuary and legally reconstitute the NAWS's management in accordance with the bylaws of this registered charity, which for years had received donations from the U.S., resulted in the first wave of opposition against IPAN and against her in person. This increased from other quarters the more she learned about other illegal and corrupt activities that short-changed tribal peoples and put wildlife at risk in the surrounding 260 square-mile Mudamalai Wildlife Sanctuary. This region has the largest remaining elephant population left on the subcontinent and also viable populations of tiger, panther and wild dogs (called dhole).

By her sheer presence, knowledge through informants, and by direct witness and video-taping of much of what is endangering this Biosphere Reserve, Deanna is, as a foreign presence, a clear threat to some— and an important ally, if not the last hope, for the many people who are coming to her now to seek her support in helping save the last of the wild in this most beautiful and relatively remote part of India. Her on-the-spot photos of an elephant, for example, who had been illegally electrocuted by a land owner's high-charged field fence, provided essential documentation to mobilize government action to better protect the endangered elephant population.

I am even more amazed at how much IPAN has accomplished since its inception in April 1996, knowing the time-consuming daily frustrations of working there, with support staff never being able to keep to reliable work hours, frequent power failures and no telephone especially during the monsoons, plus sudden shortages of kerosene and cylinder gas needed to cook our food, dogs' food, and to sterilize surgical instruments. To get to our nearest fax machine, long-distance telephone service and supply of diesel for our jeep is an hour's high-risk drive up a narrow mountain road with 36 hair-pin bends and buses and trucks coming down with gears grinding and sometimes brakes failing.

In spite of these day-to-day difficulties and the enormous animal treatment caseload that on many days is like a war-zone emergency "mash" hospital, Deanna says, "No animal is ever turned away."

I don't get to India or see my wife as often as I would like, but I am proud to be associated with IPAN. In spite of the initial obstacles and the probably inevitable local minority opposition that is now behind us, thanks to Deanna's courage and tenacity, I see IPAN as becoming a model project for other rural communities. In those rural communities located in and around wildlife preserves still rich in biodiversity and rare and endangered animal and plant species, and where there is a large population of domestic animals in need of veterinary care, providing such care not only helps improve indigenous peoples' health and wealth, it also helps protect wildlife whose numbers are periodically decimated by diseases like rabies, distemper, and foot and mouth disease that they contract from a neglected domestic animal population. It is a tragic irony that in many of the world's last remaining "hot spots" of biodiversity—regions rich in wild flora and fauna—there is a human population explosion, loss of wild lands, and excessive numbers of livestock, all compounded by human greed and corruption. We see the poor getting poorer as rich investors purchase farm and pasture land to set up guest lodges and for illegally irrigated crop production enterprises that are used to launder "black" money and as tax write-offs. The landless poor are forced to encroach on forest and wildlife preserves and engage in poaching, illegal grazing, farming and extraction of forest resources for firewood and other purposes. We have documented tribals being paid by one entrepreneur to harvest forest resources nonsustainably, destroying entire trees. This enterprise is actually subsidized by European donor agencies dedicated to sustainable development, but unaware that their moneys have been used to build a guest lodge and jeeps for eco-tourism. Corrupt land deals are made by others ostensibly to set up schools and other facilities for tribals, which serve as a cover for a variety of nefarious activities that leave the poor poorer and ultimately disenfranchised. These problems in the Nilgiris are not insurmountable, and there is much left that can and must be saved.

The greatest miracle that IPAN has accomplished to date is the following transformation. When we first came to the villages and saw the sad plight of so many sick, injured and starving animals, the people seemed to be callously indifferent. Mange-ridden and maggot-infested dogs were often kicked and stoned. But after seeing how we cared for such animals in the street from our jeep that we stock with basic veterinary supplies and equipment, we saw a dramatic change in the people. People started to take us to treat animals in need, sometimes keeping them secure until we came. Clearly, when there is no one and no resources available to care for sick and injured animals, people turn a blind eye to animal suffering. Where there is no hope, there is no compassion.

One of the most rewarding aspects of this project is to witness the wonder and appreciation in children and the poor and the elderly when their dogs, puppies, calves, cows, bullocks, goats, kids, sheep, and lambs are made well. But best of all is to be greeted by the animals themselves, especially the dogs and the old cows who show in their eyes their gratitude and trust.

The Animal Refuge

Nearby at Hill View Farm, where resident and visiting volunteer staff live, IPAN's **Animal Refuge** shelters 70 donkeys, abandoned, sick, and injured cattle, ponies, dogs, cats, and a fluctuating number of many other animals needing special long-term care prior to being returned to their owners, or original habitat. On any day IPAN may have a dog with a broken pelvis or recovering from leg amputation, an orphaned wild boar piglet, an injured monkey requiring intensive care, or a group of people standing by the Animal Refuge gate with puppies, calves, and lambs in baskets and around their shoulders or in their arms—all needing attention.

IPAN is **restoring this farm**, turning it into an oasis with an organic garden, biogas and solar or wind-power energy, which serves as a model for sustainable agriculture and resource use.

The Animal Refuge borders on the **Mudumalai Wildlife Sanctuary**. The region is home to a diversity of tribal peoples like the Todas, the "Honey Bee" and "Elephant" Kurumbas. It boasts the **largest remaining elephant population in India** and an incredible diversity of other fauna, like the Indian Guar or bison, Nilgiri langur monkey, endangered tiger, and panther. It is world-renowned for its bird life and flora; many plants are being used by tribals for various medicinal and other traditional purposes.

But for the electrified fence around the farm buildings, elephants and panthers would be breaking in every night to respectively raid IPAN's fodder store and take the dogs for their dinner. Roof rats and cobras around the buildings constantly remind staff of their close proximity to the jungle. During the monsoons when the road washes out, IPAN staff uses its bullock cart, like a covered wagon, to transport injured ponies and other large animals.

A monkey severely mauled by dogs receives veterinary attention from IPAN.

Veterinary Services

IPAN provides veterinary services to over a dozen villages and remote tribal settlements with IPAN's **24-hour mobile veterinary clinic/ambulance**, fully operational **hospital**, and permanent **animal refuge**.

IPAN also **prevents much human sickness** by controlling the spread of zoonotic diseases, by vaccinating dogs against rabies and treating them for mange (which causes scabies in humans) and internal parasites. These and other diseases that also afflict peoples' cattle, sheep, and goats, causing great animal suffering, are treated by IPAN's experienced and dedicated staff. This helps villagers and tribal communities whose livelihoods depend significantly upon the health and welfare of their herds and flocks.

IPAN staff can be found at **any time of day or night** in the jungle or in some village helping deliver a calf, performing a Caesarian operation on a cow by flashlight and firelight, or giving emergency treatment to a pony or buffalo that had been hit by a truck or attacked by a tiger, panther, or pack of dholes (wild dogs). IPAN's **veterinary emergency service** on many days is like a war-zone triage unit where staff experience battle fatigue from treating so many animals.

If you were to follow IPAN's jeep on a routine animal treatment call, or post-treatment check-up, you would see adults and children lining up with other animals—emaciated calves, sick puppies, sheep and goats with maggot- infested bite wounds—all needing treatments. Every treatment in the public eye is a lesson in **humane education**, promoting respect and concern for animals and giving hope to their owners and relief to the creatures themselves. Many animals that IPAN has successfully treated greet staff in the villages, while adults and children watch in quiet amazement and evident gratitude.

Animal Protection Laws Not Enforced

Neither the municipal (Ooty) Society for the Prevention of Cruelty to Animals, or the Collector (chief municipal authority), or the Animal Welfare Board of India, or the local police, have ever helped Deanna enforce India's Prevention of Cruelty to Animals Act. Her earlier appeals to stop the cruel and indiscriminate rounding up of dogs by the municipal authorities in Ooty in the name of rabies control, which recently killed some 350-400 dogs, including some that IPAN had vaccinated, by offering IPAN's spay-neuter and anti-rabies vaccination services for free, were ignored by all the above parties.

The latest complaint she has against the Nilgiris Animal Welfare Society (NAWS) is that the ponies she had been caring for two years before she was forced out of the NAWS 52-acre sanctuary, are now hobbled (front legs tied with ropes) to stop them from crossing the river to her new Animal Refuge that is the only place where they can get any nutritious food. They are now starving and suffering. "The lights in their eyes have gone out and no one will do anything to stop it", was Deanna's last word to me on this issue.

Education

IPAN has developed videos and other humane educational materials for school children, and other training videos to teach farmers basic animal husbandry and disease prevention. Children always gather when we treat animals in the streets where our work demonstrates the humane principles we bring to local schools.

Another important function of IPAN is providing animal welfare law enforcement training to individuals who serve as humane law enforcement agents, and to familiarize local authorities with India's Constitutional mandate and various laws concerning animal and environmental protection.

IPAN's senior veterinarian Dr. M. Sugumaran teaches other veterinarians how to spay/neuter dogs.

A Day with IPAN (in September 1997)

Like the long-running T.V. series "Mash," Deanna runs a 24-hour animal triage hospital. A typical day begins at 7:00 a.m. with local staff being instructed and supervised to care for close to 200 animals. Currently, these animals include over 60 donkeys, over 100 cattle and calves, and 14 dogs, plus anywhere from five to ten inpatient animals being held in recovery.

IPAN Director Deanna Krantz with school children.

No day is ever the same. Being prepared for the unexpected is the modus operandi: no kerosene available; power's out; two of the local staff of five don't turn up (one sick, one had a relative die); the promised "absolutely most definitely" truck load of fodder hadn't been delivered the previous evening. So someone will have to take the one and only vehicle to find out, since there's no telephone communication. This means that follow-up treatments of various animals in the surrounding villages will be delayed, we will probably miss lunch again, have dinner around ten, and turn in close to midnight. A starving village pup was picked up for treatment at IPAN's refuge.

One day when I was working at the Sanctuary in September 1997, the day began with a woman at the gate with a goat that had been mauled by dholes or wild dogs. While I was stitching up the goat, there was a phone message that a rabid dog was on the loose creating havoc in the village of Moyar. As I was finishing treating the goat, four men of the "Honey" Karumba tribe arrived. They had walked twenty miles to have someone from IPAN come and save a cow trying to give birth to a calf. We had them wait until our field assistant returned in the jeep from doing follow-up calls on animals we had treated earlier. He was always late returning because there were always new cases to see. Later I delivered the calf, who was stillborn, but at least I had saved the cow. The rabid dog, fortunately, quickly developed paralysis and died before biting anyone.

No day is more exhausting than another, treating animals infested with maggots and mange, weakened by starvation, injured by predators and careless drivers, and harmed by desperate and primitive treatments that people have adopted because no effective veterinary services or education in basic animal husbandry have ever been provided before. IPAN is changing all of this, and I found hope in the sheer number of treatments that animals received and were made well—over 2,500 in the past year.

Anti-rabies vaccinations and treatments for mange have been provided to over 700 dogs. More could be done, but typically a 2-day rabies vaccination clinic for one village can take several days to complete because of other crises: a pony hit by a truck; a cow mauled by a panther; a monkey bitten by dogs; a rabid dog in a distant village that's biting every animal and person it can reach and no one can catch it or dare get close. Another starving, mange-afflicted street puppy was saved by IPAN.

The primary economy of rural India is still based on animal draft power, manure, and milk, yet cattle received no adequate veterinary care in the Nilgiris until IPAN was established. As IPAN continues to provide veterinary services and humane education so vital to the health and welfare of community animals, local interest and support is being built up to make the Project more self-sustaining and locally managed. IPAN's spay-neuter program, widely accepted by the villagers, is already making a difference.

There is no other place in India like IPAN's Animal Refuge where there is land exclusively designated for the care and welfare of domestic animals in the heart of one of the richest wildlife regions in India, and in one of the most culturally and biologically diverse places on Earth.

The suffering of animals is intimately linked with the marginalization and poverty of indigenous peoples. Likewise the loss of cultural and biological diversity is linked with the exploitation and ultimate extinction of tribal cultures and unique wild plant and animal species and communities.

Addressing these linkages are part of IPAN, and already the unprecedented local political and social consequences are being felt, and reported in India's national free press. But what impressed me most of all, and affirmed the importance of IPAN, was seeing how sick animals, once provided proper treatment and care, responded; and how the villagers, especially the children, and I cannot mention this too often, expressed their joy and wonder when these recovered animals greeted us with such obvious gratitude and love whenever we came into their villages.

2003 Synopsis of Activities

Wildlife Protection & Habitat Conservation

IPAN's constant presence during in-field treating animals, and network of in-field informants resulted in the successful apprehension of poachers by the state of Tamil Nadu Forest Department; there were several cases of illegal land encroachment, including construction, cultivation, grazing of livestock, cutting of trees, collection of forest produce and poaching/killing of wildlife, including endangered species such as the elephant, panther or leopard, and guar (Indian bison). We have assisted the State Forest Department on several occasions with our full-time veterinarian performing autopsies on these animals and other wild species in order to determine the cause of death. For elephants, the major causes of death continue to be gun-shot wounds, electrocution, poisoning, wire snares on legs that become seriously infected, and fatal injuries to the mouth caused by home-made bombs in jack fruits.

When mother elephants are killed, adequate care of orphaned infants is not available, and although IPAN has assisted the authorities on several occasions, care will never be adequate for these infants or for other orphaned and injured wildlife, until a fully operational, well-staffed and managed Wildlife Orphanage and Rehabilitation Centre is established to serve this region, which the U.N. has designated as the Nilgiri Global Biosphere Reserve (one of only 400 world-wide, recognized for its unique biological and cultural diversity).

IPAN continues to monitor wildlife research and conservation activities, and agricultural development schemes in the area, in part because of the lack of oversight and accountability, and because of already documented problems, such as the accidental killing of a nursing mother elephant during chemical capture to have a radio collar fitted to her neck by Indian Institute of Science researchers, and the death of another adult elephant in a mud-hole caused by silting in the river following the construction of a small dam by the Forest Department to hold water for crop irrigation.

IPAN also provided transportation and field supplies for Government anti-poaching teams that were stranded in the forest without food, flashlights, rain-gear, and other basic equipment that should have been provided to them since they were being funded in part by a grant from the U.S. Fish and Wildlife Service to the State Forest Department. In spite of death threats for being involved in these and other activities and exposés, IPAN broke a ring of teak wood poachers that included one senior Forest Department employee, who was fired after IPAN revealed to the press that our informant, a low ranking Forest Department employee, had been falsely arrested after being beaten up, and had teak wood "planted" on his property. IPAN bailed him out of jail and got him medical attention.

IPAN provides the only veterinary service to perform autopsies on livestock killed by tigers, leopards and wild dogs (dholes), so that the owners can get the necessary certification in order to secure government compensation for the loss. This service essentially stops farmers from retaliating against these increasingly rare and highly endangered predators.

Our continuing contribution to wildlife conservation in the region includes the population reduction of domestic animals—dogs through spay/neuter, cattle through neutering scrub bulls and providing hybrid bulls whose offspring are better milk producers, so that the farmers will shift from keeping many cattle simply for manure production (for sale as fertilizer) to keeping a few cows for more lucrative dairy business. Many diseases like rabies, distemper, foot and mouth disease, and hemorrhagic septicemia are spread by domestic animals to the wildlife in the surrounding preserve, and IPAN is constantly monitoring the health status of the domestic animal population, vaccinating the same, alerting the authorities when an outbreak has occurred, and facilitating subsequent containment after notification.

Animal Welfare And Health

IPAN continues to press for proper veterinary treatment protocols and improvements in the handling, training and care of captive elephants and other wildlife in the region. After several years of indifference, harassment and threats of retribution from various sectors, we have been successful in engaging the police in enforcing the Prevention of Cruelty to Animals Act, and in winning several (costly and very time-consuming) court cases, especially cases of cruel handling and transportation (cattle, water buffalo and ponies). Many such animals are sheltered at IPAN s Hill View Farm Animal Refuge under court order when cases are pending, and after cases have been judged in our favor, and there is nowhere for the animals to go. Of the 400 animals at the Refuge, cared for by our team of 15 dedicated local staff, several are the ponies and mules being given permanent sanctuary at the behest of the Animal Welfare Board of India after successful court battles to secure their custody and freedom from cruelty and suffering. The resident herd of ninety donkeys (all males are neutered) was rescued from a defunct animal shelter and donkey rest home , where they were never cared for, and were being bred for the offspring to be sold into hard labor.

There is no "average day" at IPAN s headquarters at Hill View Farm, because there is always the unexpected emergency—an outbreak of rabies in a village in the heart of the wildlife preserve; a valued cow in a remote settlement needing a Caesarian operation to save her life and that of her unborn calf; an orphaned monkey that has been brought in suffering from burns from exposed electrical wires; or a dog who has been shot, hit by a vehicle or attacked by a wild boar. On average we treat some 10-15 cases every day. Some are brought to the Refuge, where owners will sometimes stay overnight with their goat, dog or calf. Other cases are in-field, often combined with routine vaccination-visits and periodic dog spay/neuter surgery (under general anesthesia) to villages and tribal settlements, where we also advise livestock keepers on good husbandry practices, proper nutrition, disease prevention and parasite control. In providing these services, and a 24-hour emergency service with our two jeep-ambulances (that are wearing out), we have earned the trust and support of the local community that has become our eyes and ears for what is going on in the jungle. This network of informants enables us to investigate, document and report illegal activities that are harming wildlife and habitat, and which the local people would not report either to the press or to the authorities, for fear of retribution, or because, as was the status quo before IPAN came to the Nilgiris, their voiced concerns would be ignored.

IPAN has thus become a voice for the indigenous people who care deeply for wildlife conservation and habitat protection, and who possess much knowledge about the bioregion, its history, the medicinal and other uses of forest plants, the sustainable use of natural resources, and especially the decline in the numbers of elephants and other endangered species, and the many conservation and development schemes in the past that have, without exception, caused more harm than good.

In summary IPAN continues to put into practice the holistic principle of One Health in the Nilgiris, which means Earth Care + Animal Care + People Care. The protection and conservation of wildlife and habitat (Earth Care) is largely dependent on providing health care and welfare for a sustainable domestic animal population (Animal Care) which in turn helps the people (People Care) who are economically dependent on their livestock, and who can suffer from diseases contracted from unhealthy domestic animals, just like the wildlife that IPAN continues to vigorously protect. The voice for the animals, the voice for the people, and the voice for Nature is one and the same. As the one voice in the Nilgiris, IPAN continues its work.

India Project (IPAN)
for Animals and Nature

The Dalai Lama's Message

THE DALAI LAMA

Today, together with a growing appreciation of the importance of human rights there is a greater awareness worldwide of the need for the protection not only of the environment, but also of animals and their rights. Unfortunately, there continue to be those who feel it is acceptable to hunt, abuse or exploit animals, resulting in their suffering and their painful deaths. This seems to contradict the general spirit of egalitarianism growing in most societies today.

I deeply believe that human beings are basically gentle by nature and I feel that not only should we maintain gentle and peaceful relation with our fellow human being but that it is also very important to extend the same kind of attitude towards the environment and the animal who naturally live in harmony with it. As a boy studying Buddhism in Tibet, I was taught the importance of a caring attitude towards others. Such a practice of non-violence applies to all sentient being—any living thing that has a mind. Where there is a mind, there are feeling such as pain, pleasure and joy. No sentient beings want pain, stead all want happiness. Since we all share these feelings at some basic level, we rational human beings have an obligation to contribute in whatever way we can to the happiness of other species and try our best to relive their fears and sufferings.

Deanna Krantz and her husband Dr. Michael Fox have been instrumental in establishing a refuge for animals in the Nilgiri Hills of South India. It is a place where animals who have worked all their lives can retire, where sick animals can be healed. This work is an inspiring example of compassion in action. I firmly believe that the more we care for the happiness of others, the greater our own sense of well-being becomes. Cultivating a close, warm-hearted feeling for others automatically puts the mind at ease. This helps remove our fears or insecurities and gives us the strength to cope with any obstacles we encounter. It is the ultimate source of success in life.

I pray not only that the help being given to animals in the Nilgiris may continue, but also that other people may be inspired to emulate this good work of setting up similar animal refuges elsewhere in the world.

October 31, 2002

Around Hill View Farm Animal Refuge

Emerging from the night through the mist and low clouds, the Blue Mountains move from indigo into purple as the first peacock cries as dawn breaks and a troop of langur monkeys whoops and barks at the panther slouching home beneath them into cool shadows where the wild boars root, cobras and hyenas pass the day, and a family of elephants walks across sunbeams going down to the river to drink and bathe. The otters are playing upstream from a spotted deer's remains that the wild dogs have just left to the flies. A shy jackal and two hungry dogs move from the waking village with its rising smoke as people boil their water and rice, clang milk churns and call their cows, a timeless mantra in the rustling stands of bamboo. As the day passes along the river, shimmering with dragonflies, the babble of water and stones becomes the babble of warblers and orioles, in the deep jungle breathing with the seasons, for eons dreaming into life the infinite, the mystery, the beauty, and the silence between the pulse of sacred drums as tribals dance into the night in Deanna's donkey corral, and two cow-dogs toast themselves beside the fire.

Following page: IMAGES of Hill View Farm Animal Refuge showing some of the resident animals, including the donkeys, cattle and horses being fed in their corrals after a day free-grazing with their handlers.

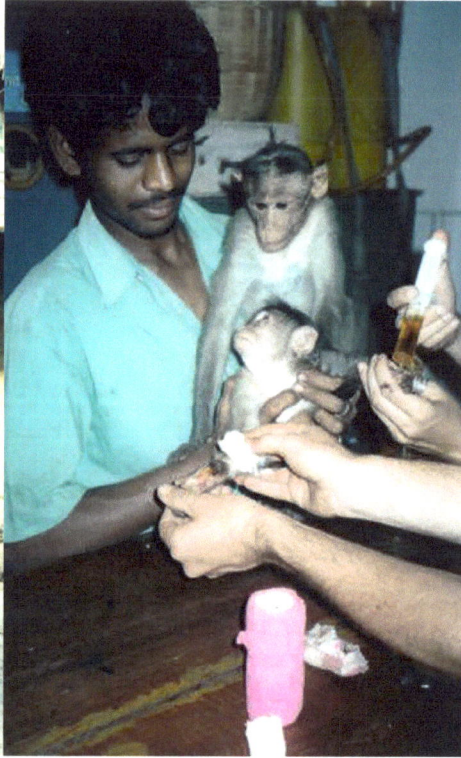

Deanna welcomes a mule, retired from the Kashmiri military conflict to her Refuge while Manya, an orphan Bonnet macaque, helps hold and comfort an infant brought in for treatment of an electrical burn.

Chapter 2
India's Dogs: Improving the Health & Welfare of the Nilgiris Aboriginal Dogs

Over 25 years ago I (Fox) went to the Nilgiris or Blue Mountains, part of the Western Ghats in Tamil Nadu, South India, to do field research on *Cuon alpinus*, the dhole or Asiatic wild dog. Rudyard Kipling once called this jungle and forest dwelling pack hunter the "red peril."

I never got as close as Mowgli, and it took me twenty years to get back to the area that UNESCO, in recognition of its unique biological and cultural diversity, has designated a precious Global Biosphere Reserve. It urgently needs what the late David Brower, founder of the Sierra Club, called CPR: Conservation, Preservation, and Restoration. The surrounding jungle, forest and shollas (open hills) are rich in flora and fauna, including the endangered Guar (Indian bison), Bengal tiger, leopard, Sloth bear, Pangolin, Nilgiri Langur monkey, and Malabar squirrel, as well as the largest remaining wild elephant population on the Indian subcontinent.

The whole ecosystem—a fragmented patchwork of state managed, but heavily exploited, wildlife preserves, reserve forests, villages, tribal settlements, plantations (tea, coffee, spices and eucalyptus)— is but a few heartbeats away from collapse and the mass extinction of indigenous species and peoples.

This is where the indigenous domesticated dogs come in as part of the problem and part of the solution. I don't mean Kipling's red peril dholes (or *Chennai* as they are called in Tamil). I am referring to the local, pariah-type "country dog," a landrace or natural, indigenous, aboriginal breed.

The earth is red, the wild dogs/dholes are red, and so are many of these beautiful and intelligent natural dogs in every village and tribal settlement, along with other coat colors (see photos below). Though some have folded ears like a terrier, most have upright ears and a conformation not unlike the Australian dingo. But, they are smaller (11-23 kg or 25-50 lb.), males being more muscular and thicker jawed than the females. Most have short, "tropical" coats, though there are variations, in part probably due to introduced European pure-breeds and crosses that rich locals keep as much for status as to guard their property. Two or three local dogs from the village would do a far better job as protectors than most of these pure-breeds, but unfortunately many people look down on the "pariah" or "country dogs" as being inferior—perhaps a sad legacy of British colonialism.

Masinagudi village dogs being fed, a morning community ritual by a caring few,
1976; note uniformity of size and diversity of colour.

Working in this area helping Deanna Krantz with her animal refuge and veterinary hospital, I have learned more about these domestic dogs, and the fate of the wild dogs and their jungle wildlife preserve, than when I came to the Nilgiris as a scientist in the 1970s to study the *"Whistling Hunter,"* as I entitled my book of field studies of the dhole (or Asiatic wild dog).

We spayed, neutered, vaccinated against rabies, distemper and parvovirus, and mange-medicated and wormed village dogs and treated them for various diseases and injuries, also educating some owners about their proper nutrition and general care. We thus helped prevent the spread of communicable diseases to people and farmed animals (notably rabies), and also parvovirus and distemper transmissible to various wildlife species—dholes, hyenas, jackals, wild cats, civet cats, "Sun " or Sloth bears, mongooses, and of course the tigers and leopards.

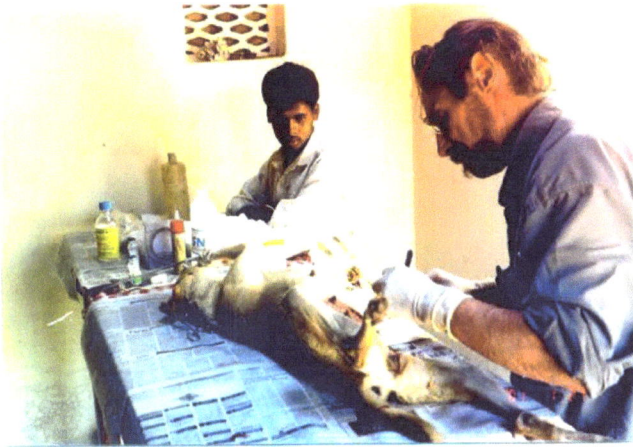

Dr. Fox spaying a dog.

As our team spayed and neutered the indigenous aboriginal dogs, Deanna and I worried that we might be hastening their demise as the offspring of purebred and crossbred dogs were preferred by most local people. So while we encouraged the spaying and neutering of these latter dogs, we also promoted the adoption of our "rehabilitated" (properly fed, socialized, wormed and vaccinated) pups from aboriginal dogs. We advised everyone to keep the purebred dogs and their often aboriginal dog-crossed offspring out of the region.

Such advocacy was not breedist or some purist ideal. On the contrary, it was to stop the terrible suffering that we saw in the Alsatians, Dobermans, Cocker spaniels, Labradors, and their progeny. They suffer terribly because they are genetically unsuited for the life and environment of the aboriginal dogs of the Nilgiris.

This is evident in their inability to thrive on the subsistence diet of local dogs, whose places they are now taking. They came to us with rickets and overwhelming infections and infestations because their immune systems were so impaired by malnutrition. This was in part because they were rarely able to scavenge and find food for themselves since, unlike most of the indigenous dogs, they were more often house and yard-confined.

These purebreds also strike us as lacking the instinctual "native" intelligence of the local dogs, especially when it comes to fighting off wild boars and leopards, looking out for heavy traffic and elephants, and avoiding snakes. Most purebred dogs we saw in the region did not have the same innate fear and alarm reactions, and their inquisitiveness, playfulness and naïve hunting behavior got them into trouble more often with snakes and other wildlife.

Our years of attention to the village dog community meant having to track down and apprehend dog-nappers who came into the villages and took off with their best dogs and puppies that we had been caring for.

Leopards enjoy eating dogs, but the aboriginal dogs are very wary and instinctively stay indoors at night, where they are always alert to the slightest noise.

With the free-running pack of over 40 aboriginal dogs in the compound at Deanna's Refuge, and the many local dogs whom we have treated for a host of conditions, we have been fortunate to come to know their various ways, and appreciate their different temperaments and personalities. Most show extreme wariness toward anything or anyone unfamiliar. They learn quickly and have incredible memories. Living with two of these dogs in Washington D.C. has been quite an education, especially seeing them relate to various purebreds. Interestingly, they have a greater affinity with American "mutts" (most from the local pound) and the occasional "natural" dog like one in Dubai and another in Thailand.

The resident pack at our Refuge was the core of a "Peaceable Kingdom" and a revelation for anyone interested in inter-species communication: Lambs playing butt-head and chase-me with the dogs; calves high-tailing it as three or four dogs chase them around the compound; as "Manya" an orphan Bonnet macaque monkey, rides on dog Shadow's back.

Shadow and orphaned Bonnet macaques play together at Deanna's refuge

"Doctor" Mani grooms other dogs and licks their wounds at the Refuge and is the most empathic care- giver.

"Aunty Xylo" plays with "Snapps", a newcomer to the refuge, assuming the role of welcoming foster- mother for incoming young dogs and pups.

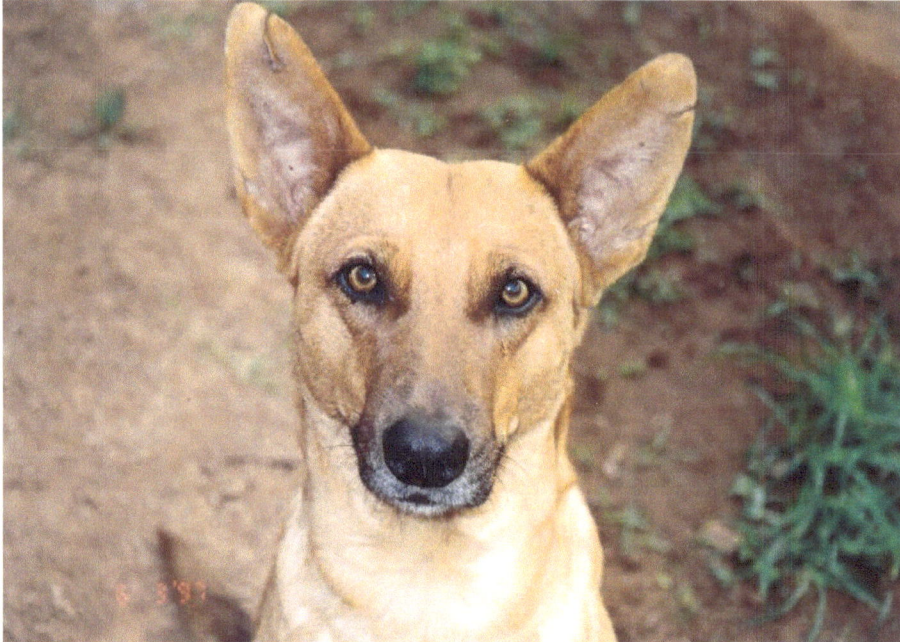

"Bingo," refuge resident, a classic "Red Dog" phenotype of the Nilgiris native/ Aboriginal dog, and a formerly prized hunting dog prior to his injuries from a wild boar.

These dogs had different barks for different occasions: Visitors at the gate, wild boars outside the fence at night, or a monkey in a tree. They would often dig up and eat dirt, a source of trace minerals. Those in the villages sometimes formed packs and hunted down spotted deer, which we discouraged, since they were often used for such illegal poaching.

In the unique environment of our Refuge, where unlike the usual village environment, the dogs do not have to compete for food and all are neutered and spayed, aspects of their behavior and personalities surface that gave me a deeper appreciation for the aboriginal, native dog of the Nilgiris. The smallest of our males, Dean, was the pack leader. Charisma and not simply size and strength evidently determined who would be "top" dog. Dean had a hard start in life, being semi-starved and left untreated for mange and worms for months. We rescued him from a defunct animal shelter in time to nurse him through distemper.

There was "Mani," another red dog who came to us with a shattered pelvis. He was the "caretaker" of the pack, checking the dogs over for ticks and grooming them with tender care. "Bingo," a classic red dog, was one of the brightest and best. We amputated one of his front legs after he was brought in mauled by a wild boar while out in the jungle with his owner, who gave him up to us.

"Xylo," a brindle female who was severely malnourished as a pup but somehow survived on the streets for the first 5 months of her life (and eventually lived with us in the US), was the "auntie" to all new pups brought to the refuge, gently licking them and encouraging them to play.

"Shadow" was an unusually large white village dog who dragged himself to the Refuge gate after having his back broken by a vehicle over a mile away in the village of Mavanhalla. He had never been to the Refuge before yet knew where to come for help—like the water buffalo who came one day and stood by our gate and waited to be treated for a horrendous maggot (screw worm) infestation.

Wounds quickly become maggot-infested and the maggots slowly eat the animals alive. Dogs are often victims of the "screw worm" fly maggots that cause more agony than the mange, eating holes into their throats, heads and sides after they sustain the slightest wound. Chronically malnourished dogs have very weak skin that tears easily and heals slowly. Our spay-neuter program in the villages greatly reduced fighting and the horrendous incidence of secondary maggot infestation.

Dean at 4-months, suffering from starvation, distemper and mange,
and as a 4-year old recovered adult.

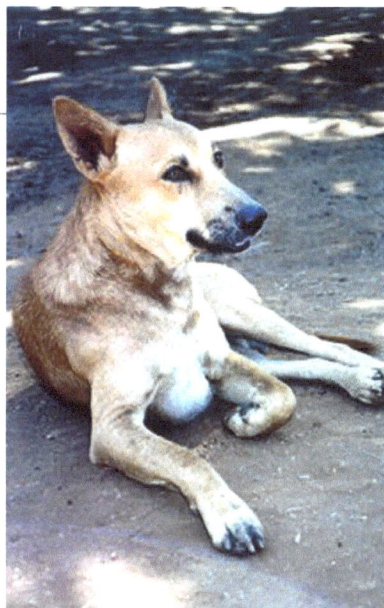

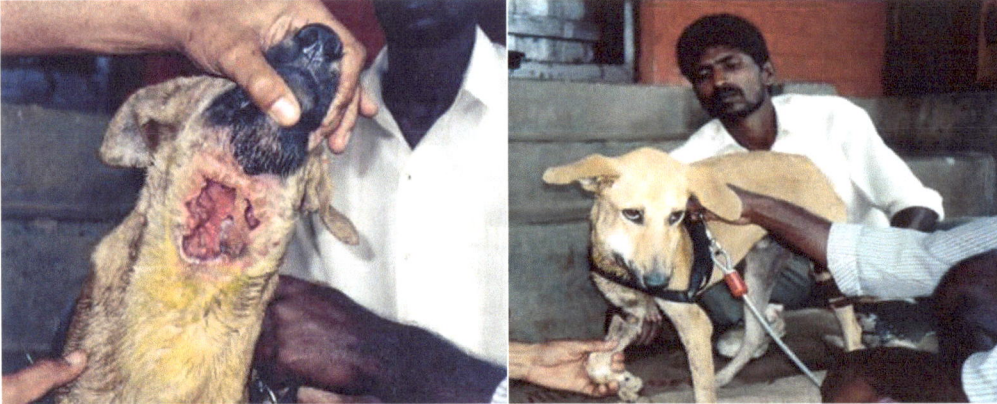

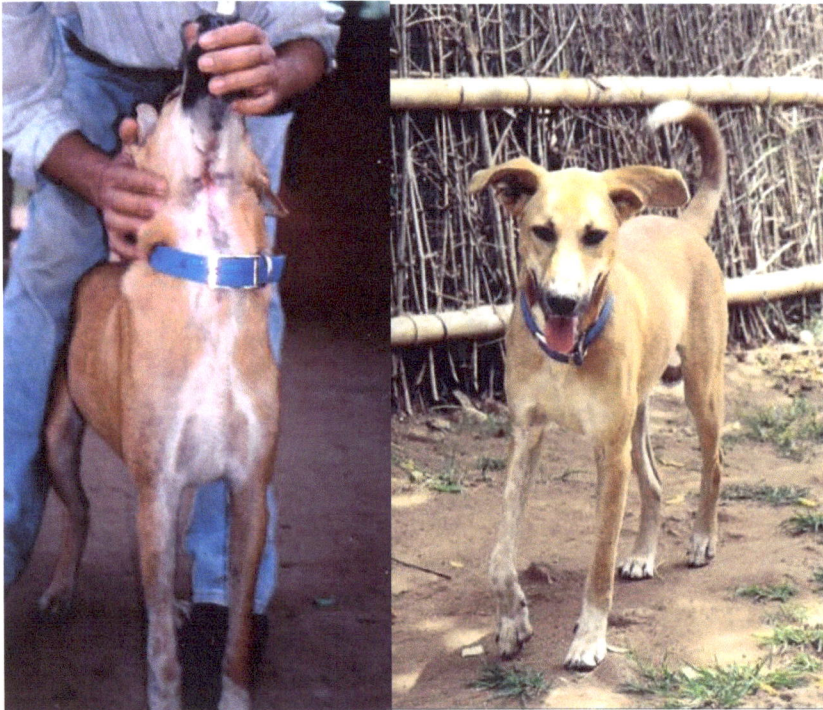

"Jockey" was brought to the Refuge with a maggot infested throat, from a dog-bite, so severe that swallowed food came out of the infested lesion, but with IPAN's care made a complete recovery.

Starving mother nurses pups who have a poor start on life, few surviving,
a solution solved by spay/neuter.

Our staff worked round the clock to improve the health and welfare of the aboriginal dogs and other domestic animals, which benefits the people and helps protect the last of the wild. After a long day in the field, I always enjoyed returning to the Animal Refuge to be greeted by the resident dog pack like a long-lost friend.

Canine 'love pyramid' greets Dr. Fox while holding
pack leader Dean: Deanna's animal refuge, S.India

Dean, leader of the pack at IPAN's Refuge.

Some Of IPAN's Treatment Successes

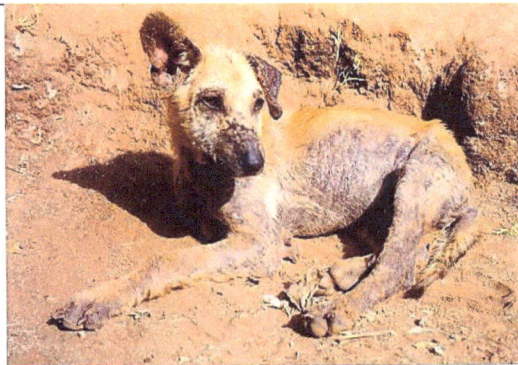

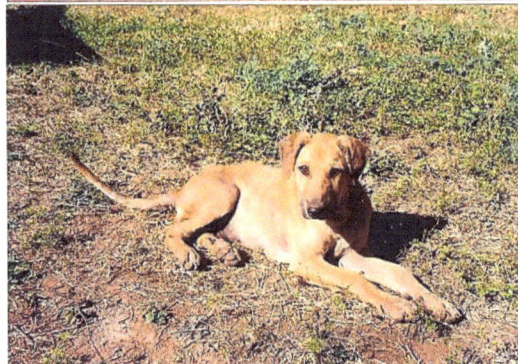

"Snapps" was caught and brought to the Refuge and treated for mange, worms, and severe malnutrition, showing complete recovery after 6 weeks of intensive care. "Twilight" (below) took 4- months to recover.

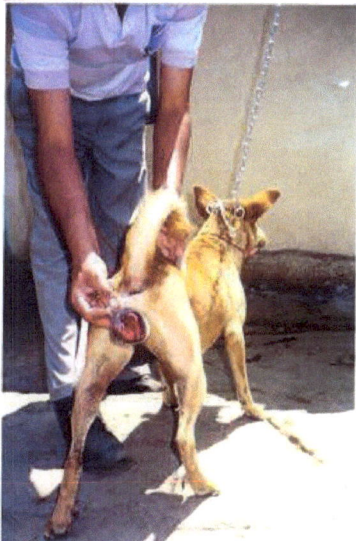

Village dog above was caught and treated and the mammary tumor removed,
while female below, one of many treated for transmissible venereal tumors.

IPAN's staff halt municipal truck and remove owned and community dogs, some already killed, in periodic depopulation roundups and mass poisoning and killing.

India street-dog snaring and killing injecting with cyanide. The dogs scream as children run or stare

Neutering prevents temporary pack formation around a female in heat and reduces incidence of dog fights and serious maggot infestations of wounds, which can mean prolonged suffering and a slow death for many dogs.

Local Children And Their Dogs

The Nilgiris Native Dog: More Insights
(With Drs. M. Sugumaran & B. J. Sugumaran, veterinarians with IPAN/USA)

In the UNESCO designated Nilgiris Biosphere Reserve in Tamil Nadu, South India, the region is noted as a unique place for its remarkable cultural and natural biodiversity. This includes the not widely known native aboriginal dog commonly known as 'Country' dog or local 'pariah.' These natural, aboriginal dogs are, however, declining in population. This is mainly due to cross breeding and fashion of passion for foreign breeds.

We conducted a survey of indigenous people's attitudes and beliefs toward their native dogs during April and May of 2009.

The Nilgiris Native dogs vary in size from 25-50lb, many adults being undersized and underweight due to chronic malnutrition. They are long of limb and tail, with usually erect or semi-erect ears. Normally all these dogs are protective and very faithful to their owners. These dogs have a good musculature, and the males are clearly more robust and have more powerful jaws than the females. All have characteristically small paws relative to their size compared to most modern breeds. Their sense of smell and tracking abilities are considered superior to other breeds. They are skilled hunters. Around other domestic animals with which they normally live in villages and Tribal settlements, like chickens, calves and goats, they are gentle and even protective, most probably a result of selective breeding and training.

These dogs have great stamina and have better resistance to many diseases compared to imported breeds and cross-breeds. They are able to sustain themselves as scavengers, often existing on a subsistence diet that for other dogs would often mean rickets, stunted growth and other deficiency diseases. They show innate nutritional wisdom, often being seen eating mineral-rich dirt and enzyme, bacteria and protein-rich feces of suckling calves.

The females are more protective towards theirs puppies than non-native breeds, and will chose to whelp in a secluded place and may sometimes burrow a den. They will often nurse their pups for several weeks longer than other dogs, pups continuing to be accepted as old as 4-5 months of age.

Tribal people rear these dogs to guide them in the forest and to hunt smaller animals. These dogs also are instinctively alert to the scent tracks of potentially dangerous panthers and tigers, wild boar and cobras, and are especially on the alert after dark. They are noted for their courage and tenacity, defending their owners from wild boar and sloth bear attacks.

These dogs vary considerably in their vocal repertoire, which is generally rich and subtle in terms of sound combinations (like growl-whines, yelp-barks and pant-huffs) that give clear indication of animals' emotional state and intentions. Some give low 'huffs' and growls when sensing danger, while others give full voice, which is not preferred when in the potentially dangerous jungle where silence is desired. They will give different barks when alerting to wild boar in the bush versus monkeys in trees, and will engage in coyote-like yip-yap howls when they sing in choral groups. One distinctive sound that some make in greeting is a coo-like twitter sound with high notes that sound like whistling, much like the whistle-call of the Dhole or Asiatic Wild dog.

The Nilgiris Native dogs come in different colors that include black, red, tan, white, piebald, and brindle. The most characteristic coat color is red, possible a parallel or convergent adaptive coloration seen in the indigenous Wild dog (*Cuon alpinus*) also known as the Dhole or Chennai, one of the few wild canid species that hunts in packs.

People sometimes favor particular dogs based on their coat color. They say that black color is a good luck to their star (horoscope) so they rear it. And there is the common belief that black dogs can see or sense ghosts or evil spirits and signal by louder barking to caution the people in the house of the danger.

The red dogs are chosen because of their great musculature and good shiny coats. These dogs have excellent temperaments and are noted for their loyalty. They are believed to be technically proficient hunters. They will even hunt larger animals like sambar deer and adult spotted deer as a pack (a traditional activity that is now illegal).

The white dogs are considered to be good luck to their owners, and they are good looking. These dogs are less vocal, being considered softer or silent compared to other color types of native dog. They are more prone to develop skin problems however, including photo-sensitive dermatitis on the muzzle especially, compared to dogs of other colors.

The piebald is cross of white and black dogs and they are also usually gentle-natured; sometimes these dogs have long hair coat. The piebald puppies are mostly loved by children as they grow as they are most affectionate to them. Black and white piebald dogs are sentimentally good luck.

The brindle colored dogs are believed to be more aggressive, but this may be because they are mostly not touched and petted by people or children due to unattractiveness, so they become timid or ferocious.

Tails are normally long and straight and are curled upward or downward in display, but some dogs have more permanently up-and curled tails. People believe that too much curling in the tail is not good and could be a bad omen so they dock the tails of those puppies with curling tails. They believe if the tail touches the back of the body it is not good for the head of the family.

These Nilgiris native dogs are in state of potential extinction due to dog breeders introducing foreign 'exotic' European breeds that are seen as a status symbol. Many of these purebreds are deliberately crossed with the indigenous Nilgiris native dog, in part to help them adapt better to local conditions, but further diluting and 'contaminating' the genetic lineage of these wonderful indigenous dogs. Spay/neuter 'birth control' programs have further reduced their numbers.

A policy decision to not neuter classic phenotypes of this now threatened domestic variety of dog would be a wise move in this bioregion.

The outside breeds, those contaminating the gene-pool of this native lineage, are German Shepherd Dog (Alsatian), Doberman, Labrador retriever, Rottweiler, Terriers, and Hounds brought in decades ago by British people and more recently by more affluent citizens.

The red dogs are predominant of those colors mentioned above. We would like to conserve the Nilgiris Native dogs due to their beauty and temperament. We also believe that this native, aboriginal variety of dog may well be the classic prototype of the earliest domesticated dog and belongs to that ancient lineage.

We continue to provide full veterinary services for these privately owned and community-living dogs, including spay/neuter, vaccinations, treatments for mange and other parasitic and infectious and contagious diseases and emergency services for injuries and acute conditions including poisoning and snake-bites.

Method Of Data Collection

We conducted our attitude survey in an informal way with the village people and noted down their remarks after visiting each of 18 villages around the Mudamalai Wildlife Sanctuary. We did not take any notes in front of the villagers to avoid exaggerated words. Many of the interviewees felt that coat color made no difference to temperament, but overall we found there was a general consensus that did associate coat color with temperament as well as revealing some beliefs and superstitions of anthropological interest.

In Each village a group of elders and youngsters participated. Though the exact number of participants are not counted, at least 10 persons were present in each meeting. We collected data from 54 places with an average of three places per village. Additional data was also collected during our visits for emergency veterinary treatments.

Native aboriginal dogs are either part of the community and free-roaming,
or live in homes as part of the family and are often ceremonially laid-out and
mourned when they die, and ritually buried.

Photodocumentation Of Indigenous Dogs

Photos below by Drs. Sugumaran showing classic red dog type, then variations in coat colour and size, with the last two photos showing two Country dog X Dobermans and a Country dog X German Shepherd/Alsatian.

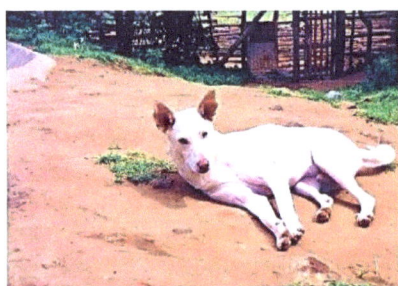

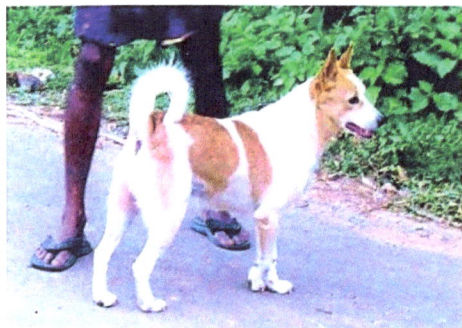

Addendum: Dogs Of High Regard: The Superdog

People of high regard in Australian aboriginal society were seen as possessing super-human powers and were the leaders, elder-advisors, and healers of the soul, saying "Dingo make us human". In my studies of dogs and their feral and wild relatives like the wolf and the Asiatic dhole, I have seen, in these socially evolved species, individuals of high regard in terms of their role in the pack as leader, arbiter and healer of conflicts; gentle and protective of the young; catalyst for play and song, and brilliant in deflecting confrontation when challenged or deceived.

These super-dogs, arguably the most highly evolved members of their race, possess many virtues all too often lacking in our own species, notably loyalty, honesty, trustworthiness, obedience, courage, devotion, protectiveness, and empathy. They can be aloof and seem distant at times, showing a reflective inwardness that is embodied with a power of presence that makes them shine! They are so observant and understanding of our behavior and intentions that they seem almost clairvoyant—mind readers indeed. These dogs, like their wolf cousins, are the natural alpha members of any social group, and they dispel the erroneous notion, still practiced by some dog trainers, that to be an alpha means to dominate through intimidation, manipulation, coercion and even physical force.

Dogs of high regard, like the dog-headed Egyptian god Anubis, enter other realms beyond our normal conscious state and awareness. For example, dogs can know when their human companions are coming home, and when a loved one away from home has died. These documented abilities mean that unlike most humans, except those like the Australian aboriginals of High Regard, dogs can enter what I call the empathosphere (visit www.drfoxvet.com and *Animals & Nature First* at Amazon.com for more details).

Dean

Dean was one dog of High Regard whom we saved as a pup from canine distemper, sarcoptic mange, internal parasites galore, starvation, and gross neglect at a defunct animal shelter in India. Perhaps it was the close bond with us that helped Dean establish himself immediately as the alpha dog. But that alone would never have helped him when challenged by stronger dogs when we were not there. Dean was one of our smallest dogs, and was weakened by the ravages of distemper when we found him at the shelter. On occasion he had seizures. Whenever he was seriously challenged, he would remain stiff and still, eyes glaring and teeth bared. He rarely needed to growl or snap.

Dean showed prescience on many occasions. He always knew well ahead of time when Deanna would be coming back in the jeep to the animal refuge. He would be the first at the gate before anyone could hear her vehicle.

He was the first to approach a new dog we had brought in to the group after quarantining, as the other thirty or so residents circled or sat while he examined the newcomer. He would police all interactions with the newcomer with calm and deliberated attention. Was he cool!

The first to alert the pack and the rest of us was usually Dean, when a stray animal was outside the refuge, or a wild boar or elephant was close. He had a different bark when there was a band of langur or Bonnet macaque monkeys in a nearby tree.

Dean, who looked like a small dingo, stocky, with sparse wiry fur, but weighing only a little over 30 pounds, was a great yowl-talker like all of these aboriginal dogs not yet too crossed out by European and other mal-introduced breeds. He had one particular whine-yowl that he used to call me out from breakfast to have me play my shakuhachi flute, which he just loved.

Dr. Fox plays to discerning audience and chorus.

On hearing my first note he would utter a cascade of high pitched whines and whistles that ignited the entire pack in a cacophony of yips, yaps, yowls, howls, and bird-like whistles and trills of happiness that spread to the nearby village where the community dogs began to sing. I imagined a wave of singing dogs passing from village to village, a tsunami of joy spreading as fast as the speed of sound through the thin morning air from our animal refuge where no dogs were ever caged except for medical reasons, and all were safe and well fed. Dean was the dean of morning song in the Nilgiris, where I pray his noble
lineage will not become extinct.

At evening time I would often find Dean away from his work-day sentinel posts around the refuge, lying and facing the setting sun. Sometimes he would be sitting on his haunches and would turn to look at me as I approached, and then return his gaze to the changing of the light.

On one occasion he bemused us all by lying in an unusual spot all day while a dowser and hydrologist tried to find the best location to drill a well. Toward day's end they gave up until Deanna insisted that they drill where Dean had been lying all day watching the workers. They eventually found sweet water deep down beneath him.

CHAPTER 3
India's Holy Cow: Problems & Solutions

IPAN

Introduction

India has the largest concentration of livestock in the world, (250-300 million cattle, 60 million water buffalo, 120 million goats, and 40 million sheep), having one-third of the world's cattle on approximately 3 percent of the world's land area. India is the world's second largest milk producer, with over half its milk coming from buffalo. Seventy-six percent of Indian people are rural, living in some 600,000 villages. The economic and social values of cattle are so great that cattle have long been seen as religious symbols and are regarded as sacred.

Cattle are vital to the rural economy, for ploughing, draft work and milk for families

In India's villages today, one can see the close relationship between cattle and their owners who have high regard for their animals as individuals, as vital family-providers, if not also actual family members. Hence the strong resistance to killing and eating such close animal allies. But this symbiotic alliance is breaking down as larger modern dairies are established and animals' individualities are lost, and as venture capitalists purchase bullocks and carts to be rented out, or leased to individuals who are complete strangers to the animals, and who have no emotional or economic interest in them.

India's "white revolution" began in 1970, a nationwide dairy cooperative scheme called "Operation Flood" that was initiated to increase milk production and to help the poor with low interest loans for purchasing milk cows. The World Bank and the World Food Program provided most of the funds, but this scheme has caused many problems (Crolty, 1980). Less grain and lands are available to feed people since more are diverted to feed dairy cattle owned by the rich. Also, fodder prices have increased, creating difficulties for poorer cattle owners and landless cattle owners.

But now, according to Prof. Ram Kumar of the India Veterinary Council (personal communication), there is only sufficient feed for sixty percent of India's cattle population. This means that of an estimated 300 million calves, bulls, and bullocks, some 120 million of these animals, especially in arid regions, and elsewhere during the dry season and droughts when fodder is scarce, are either starving or chronically malnourished.

Large numbers of 'scrub' cattle kept as manure-producers destroy wildlife habitat and compete for water.

While Moslem, Christian, Sikh, and other Indians eat meat (buffalo, sheep, and goats, whose slaughter is permitted) the majority of Indians are Hindus, for most of whom the killing of cattle and eating of beef is unthinkable because this species is regarded as the most sacred of all creatures.

Cow Worship

Cow and bull worship was a common practice in many parts of the world, beginning in Mesopotamia around 6,000 B.C. and spreading to Northwestern India with the invasion of the Indus Valley in the second millennium B.C. by Aryan nomadic pastoralists who established the Vedic religion. What is remarkable is that such worship has persisted uniquely in India to the present day. Lodrick (1981) concludes that revulsion against sacrifice, the economic usefulness of cattle and religious symbolism (especially as the Mother-provider) were factors contributing to the formulation of the sacred cow doctrine, but it was *ahimsa* (the principle of non-violence/non-harming) that provided the moral and ethical compulsion for the doctrine's widespread acceptance in later Indian religious thought and social behavior.

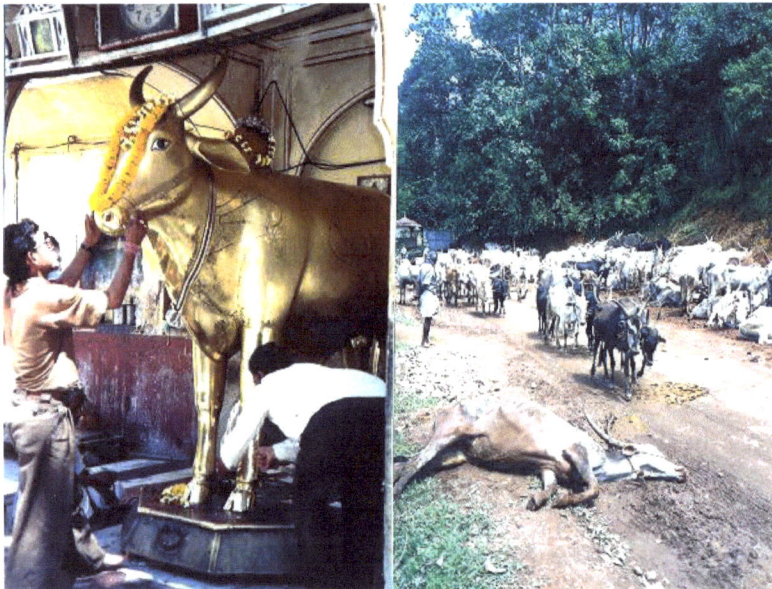

Reverence for the sacred icon of a cow does not always translate into humane treatment as per this pregnant cow collapsing on a forced march to slaughter

India can be seen as two nations in one: A majority of Hindus, for whom vegetarianism is linked to caste and ritual purity; and the meat-eating Moslems, who are seen as unclean and their touch polluting (Simoons, 1961). Moslems regard Hindu worship of temple images as heathen and immoral and their democratic views contrast with the caste system of Hindus. The Hindu elite abstain from eating meat. From an ecological viewpoint and an economic one, Hindus and Moslems are highly complementary when it comes to cattle. One eats the male calves while the other takes the calves' milk.

Cow protection has become a highly politicized core of the Hindu religion. What was once a compassionate, symbiotic human-animal bond linked with virtuous behavior (personal purity) that brought with it such principles as *ahimsa* and vegetarianism for Hindus, and for Moslems the ritual codes of animal sacrifice that helped affirm community and family ties, has now come to serve abstract economic and political ends.

Religious beliefs that ultimately contradict nature's reality and which see the nature of other creatures as unclean or immoral, become life-negating rather than life-affirming, and cause great harm (Fox 1996).

Cattle Welfare Concerns

Because of a seasonal and regional lack of fodder (and water), and because of overstocking and overgrazing, many cattle suffer from chronic malnutrition. This in turn weakens their immune systems and makes them susceptible to parasitic infestations and other diseases. Large numbers of poorly nourished cattle create a potent medium for outbreaks of infectious diseases, which necessitate costly vaccinations, which are too often ineffectual due to inadequate refrigeration.

There is also the widespread belief that there is no real cattle surplus, and that India would do better with even more cattle because their organic manure is so valuable to agriculture. The environmental damage in some regions from overgrazing is especially caused by "scrub" cattle that are kept simply as manure-makers before they are driven to slaughter or die. Their sad existence in semi-starvation, often also chronically sick, along with the free-roaming ones in villages and towns, will continue without mass public education and government assistance.

Free-ranging calves, severely malnourished, survive on garbage.

The overall cattle population must be reduced; and health and productivity enhanced through genetic improvement, and by better nutrition by establishing emergency fodder banks and sources of water to see them through the dry seasons; and alternative sources of income provided for farmers who are reliant upon cattle manure as a major product, as by raising milk-goats and producing more fodder.

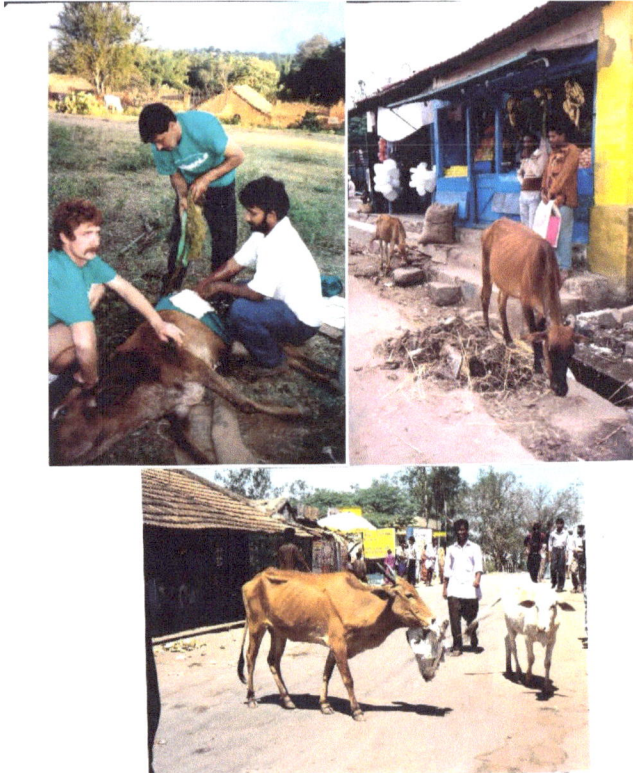

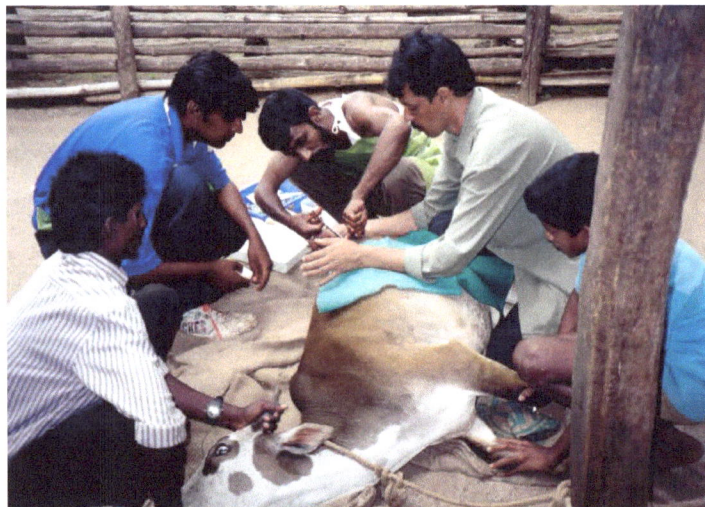

Calves and cows eating garbage become impacted with plastic bags, 10kg (above) & 20 kg (below) being surgically removed by IPAN's veterinary team.

According to *India Today* (January 11, 1996), "As long as 1955, an expert committee on cattle said in its report: 'The scientific development of cattle means the culling of useless animals ... by banning slaughter ... the worthless animals will multiply and deprive the more productive animals of any chance of development.'"

Paul Shepard (1996) criticizes one anthropologist who wrote a long article defending the sacred cow on 'ecological' grounds as a consumer of weeds and plant materials that otherwise went to waste, because this view of the sacred cow is a flagrant but familiar abuse of the concept of ecology as maximum use instead of a complex, stable, bio-centric community.

Seeing the increasing desertification of pasture lands caused by overgrazing, and cattle having less and less grazing land as good land is put under cultivation, environmentalist Valmik Thapar foresees that if the cattle problem is not soon corrected, "Finally there will be a clash because the land mass of the country can't sustain the growing human and animal population. Then the question will arise as to who is going to eat. Man or cow?" (*India Today*, January 11, 1996)

Cattle Shelters

The first animal shelters in India began with the advent of Buddhism, to whom King Ashoka (269-232 BC) converted. Ashoka ruled over much of the Indian subcontinent, converting millions to accept Buddhism, and was the first to set up shelters and animal hospitals, although some historians believe that Buddha himself was the first to do so. Ashoka put compassion into action, by caring for animals in need, and into the law also, setting up wildlife preserves and punishments for those who abused and killed animals.

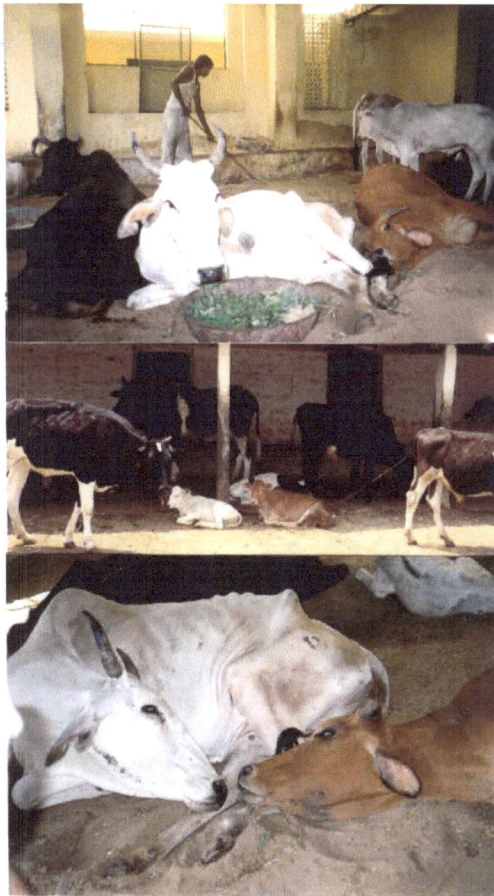

Some cow shelters, like this one in Jaipur, are well run, providing care and comfort for dying cows and bullocks.

Gowshalas and pinjrapoles are located throughout India and are supported by taxes and charitable donations from the business community. Gowshalas are refuges for cattle, often linked with the Hindu cult of Krishna, while pinjrapoles serve as a refuge for a more diverse animal population, including birds, other wild animals, and even insects and microorganisms in collected piles of household dust. A 1955 government census found there were 3,000 animal shelters maintaining some 600,000 cattle and thousands of other animals from deer to dogs and camels to cats.

Even though Indians know that the buffalo is a better quality milk producer than most varieties of cows, buffaloes are rarely found in gowshalas because they are considered unclean and not worthy of the same respect as cows.

The Gowshala Development Scheme implemented in the 1957-1961 five-year plan to provide subsidies to improve existing gowshalas were more successful during some periods than others since their implementation.

Starving, emaciated cattle at this shelter with sacks in public view giving the impression they get plenty of food but in fact the sacks were stuffed with inedible straw and wood shavings.

The prevailing view that such a fate as starvation is better than having cattle defiled by the butcher's knife, does little to encourage local public support. Levying a tax on milk, hides, manure, bone and meat meal fertilizer, and taking a percent of the profits from wholesalers of these cattle products to help defray the costs of running a gowshala that serves the community, is the kind of initiative that is needed, but which politics in many regions would preclude (Bone meal from urban cattle who live in high density traffic areas, where leaded gasoline is used, becomes potentially toxic with accumulated lead).

According to Lodrick's study (1981),all gowshalas that keep dry cows and cattle that cannot be rehabilitated for draught work, operate at a deficit. Attempts to make them more productive are not likely to significantly reduce this deficit and so without adequate community and government funding, as is the case throughout much of India, cattle suffer a fate surely worse than the butcher's knife.

The antipathy toward cattle slaughter can have absurd and cruel consequences. For example, according to the *Indian Express* (Coimbatore, February 25, 1997), local authorities "tied up a huge wild bull on the rampage." It was decided to auction off the creature for slaughter, which fetched much opposition from the devout. Someone killed the bull with some poison during the night to "save it from being defiled by the butcher's knife."

In spite of the excellent research, scholarship, and dedicated field work visiting animal shelters throughout his homeland, Lodrick says nothing about the suffering of cattle in gowshalas or of other species in pinjrapoles. Lodrick sees, in spite of their economic inefficiencies, gowshalas and pinjrapoles persisting in India because cows are held to be sacred and because of the principle of *ahimsa* that prohibits killing, even for humane reasons. This prohibition is motivated less by compassion than by the belief that to kill is to make oneself spiritually impure.

Cattle Death Drives

Millions of old, spent cows, exhausted bullocks, and young male calves are driven on foot up to 300 miles, or are crammed into trucks for transit into Kerala, or in railroad cars to West Bengal, the two states where cattle slaughter is legal. Their often bleeding, worn down hooves make hardly any sound as they pass by. Veterinarian Dr. Ghanshyam Sharma from Sikkim, in the Northeast of India where cow slaughter is also legal, sees cattle coming in from Jamma, Kashmir, Bihar, and Nepal. He observes, "Often entire hooves of these animals are snuffed out and gunny bags are tied around the wounded stumps and this way they walk." Many sustain injuries being loaded and off-loaded during part of the journey or die in transit. Some collapse on the way, are beaten, and even have salt and hot chilies rubbed into their eyes and have their tails hammered, twisted, and broken to make them get up and keep walking. Some of those being transported get trampled and suffocate, or have an eye gouged out by another's horn. Water and fodder are rarely provided during their long journeys, and even at rest stops. An estimated one million cattle are taken every year into Kerala from other southern states to be slaughtered (*India Today*, January 11, 1996).

Often cherished cattle finish up at auctions and then are force-marched or transported to slaughter.

Journalist Subhashini Raghavan, in his expose of these cattle death marches, found a complex network of middlemen traders, "who are calloused by constant exposure to cruelty" and they develop the attitude that "if an animal is slotted for slaughter, it ceases to be a living being with pain, hunger and terror." Raghavan found that vast numbers of cattle are made to walk hundreds of miles through pedestrian side-roads to escape checkpoints, on route to regional markets from local markets and then on to transfer points where they may then be put into trucks. He concludes his article stating that, "throughout the length and breadth of this birthplace of Ahimsa, the tragic march of the condemned continues unabated—a poignant symbol of our callousness, in even denying the last comforts and dignity of those who lived their lives serving us" (*The Hindu*, April 16, 1995). Cattle shelters—gowshalas and pinjrapoles—cannot possibly absorb all the unwanted cows, calves, and bullocks, since the cattle population is constantly increasing because a cow must have a calf to produce milk. The ecological damage of overstocking, overgrazing, and of millions of low-yielding milk cows and "manure" cattle is turning some parts of India into desert, devoid of trees, topsoil, and wildlife. India's 40 million sheep, 120 million goats, 60 million buffalo, and expanding human population, now estimated at 930 million, further compound this environmental devastation.

The weekly cattle 'death-march' from Tamil Nadu auction, with collection points in Karnataka to Cochin, Kerala.

Cattle with tails broken from being twisted and beaten to make them get up and move into overcrowded trucks.

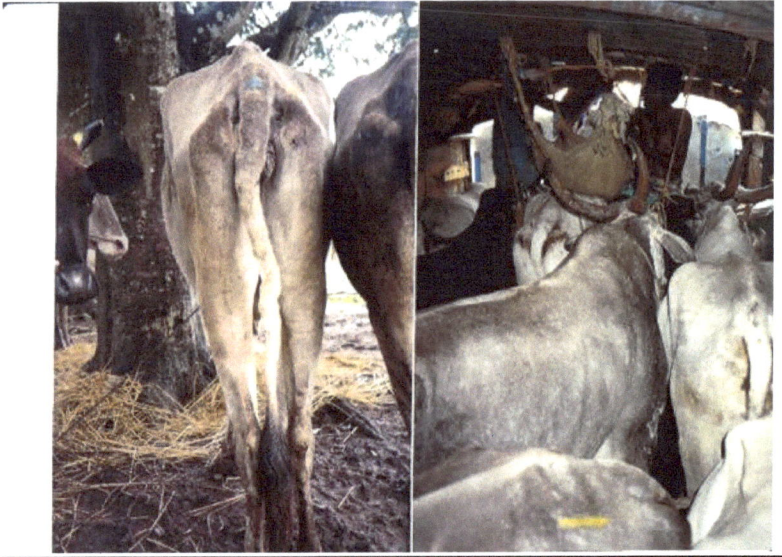

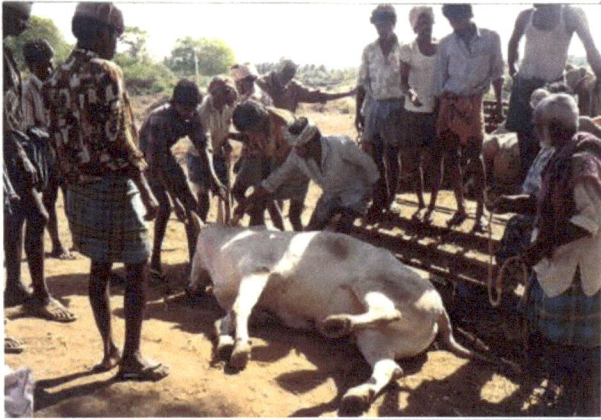

Cattle Slaughter

Belief in *ahimsa* (not harming) and in *aghnya* (not killing) possibly arose as a reaction against the Vedic religion and social order that sanctified animal slaughter, the Brahmins being the highest priestly order in the Hindu caste system that supervised the killing according to Harris, (1991).

Between the eighth and sixth centuries B.C. a new wave of philosophical treatises emerged that included references to ahimsa, and also reincarnation and karma, that were not included in the Vedas. These treatises, along with the emergence of the religious traditions Buddhism and Jainism that espoused ahimsa, were a challenge to orthodox Hinduism and may have led to the Brahmins prohibiting cow slaughter and promoting ahimsa. Yet still today thousands of animals -- buffalo, sheep, and goats especially -- are slaughtered in Hindu temples.

Cattle arriving exhausted at slaughter house in Cochin had neither food nor water on their long journey or on arrival.

Except in West Bengal and Kerala, where cattle slaughter is permitted, the Cow Slaughter Act prohibits the killing of cattle less than 16 years of age. The penalty for illegal slaughter of cattle is rigorous imprisonment for two years and a fine. Article 48 of the Constitution of India, Part IV, Directive Principles of State Policy, Article 48--Organization of Agriculture and Animal Husbandry, says: "The State shall endeavor to organize agriculture and animal husbandry on modern and scientific lines and shall, in particular, take steps for preserving and improving the breeds and prohibiting the slaughter, of cows and calves and other milch and draught cattle."

According to one government study, 50 percent of goat and sheep slaughtering and 70 percent of large animal slaughtering is illegal, taking place in clandestine facilities where there is no supervision of hygiene, animal welfare, or meat safety inspection (Report of the Expert Committee, 1987).

Of the 3,600 licensed abattoirs in India, only two are mechanized and hygienic, and these are facing strong public opposition (*India Today*, January 11, 1996).

Ritual slaughter (throat-cutting with no pre-stunning) in Cochin leaves conscious animals bleeding to death while other animals waiting to be killed on the filthy floor look on.

Article 51-A (g) of the Constitution of India states, "It shall be the fundamental duty of every citizen of India to protect and improve the natural environment ... and to have compassion for all living creatures." This is not in keeping with the predominantly religious sentiment that interprets compassion for living creatures as "rescuing" cows and other abandoned cattle from slaughter and putting them into death camps where they starve to death or die slowly from infections and parasites.

Euthanasia of suffering animals, according to the Prevention of Cruelty to Animals Act, is allowed if "it would be cruel to keep the animal alive" but only if the court, other suitable persons or police officers above the rank of the constable concur. Because of the religious opposition to euthanasia, even of dying animals in severe pain, there is no legal requirement that the owner of such an animal should have it killed. Many orthodox Hindus and Jains oppose the killing of animals for any reason because they feel it is wrong to interfere in any way with another's karma or destiny. It would seem that the doctrine of *ahimsa* as it relates to the treatment of cattle has been corrupted to serve the interests of social status, caste distinctions and politics, since lower Hindu castes, tribal peoples and others do kill and consume cattle and other animals, be they healthy or in a condition that calls for immediate euthanasia.

Indians have reasoned with me that killing a sick cow is like killing your own mother and that is unthinkable (see also Simoons, 1961).

The Animal Welfare Board of India, the chronically understaffed and under-funded government agency with limited powers to enforce animal protection laws, does help subsidize local Blue Cross and SPCA animal shelters and hospitals. But, without more support from the central government and from foreign animal protection organizations, the plight of India's animals will worsen as the human population increases and resources become ever more scarce and costly.

Vegetarianism, Religion and Politics

Vegetarianism in India, like ahimsa, has as much, if not more, to do with concerns about reincarnation, one's personal degree of spiritual purity, and place in society (caste) than with immediate concern for animals. But it is not total vegetarianism, since dairy products are consumed by most Hindus and Jains. Few are pure vegan (eating no animal products). Some Jains have agreed with me that to be consistent with their religious beliefs and with the ecological and economic dictates of the current situation, veganism is an ethical imperative. Abstaining from all dairy products would be more consistent with the principle of *ahimsa* that they hold so dear, than "saving" spent dairy cows, calves and bullocks from slaughter and condemning them to slow death by starvation in gowshalas or pinjrapoles.

Yet it is in Jainism that the principle of *ahimsa* was first espoused, most notably by Mahavira (599-527 BC), a contemporary of Buddha, although earlier Jain leaders (tirthankaras) well before the time of Buddha, like Parsvanatha (circa 840 BC), renounced the world and established an ascetic community that practiced *ahimsa*. Some contemporary Jains get around the problem of *ahimsa* by becoming land owners and having others do the farming, clearing the land and killing wild creatures, ploughing the land and killing worms, and using all manner of pesticides.

Jainism reached its peak between the 5th to 13th centuries AD, spreading across much of India, then was superseded by Hinduism, and then by Islam following the invasion of the subcontinent by the Moguls in the 11th century. Moslems killed and ate cattle, which was anathema to the non-tribal, upper castes of Hindu society. Cow protection and worship then gained political importance and popularity in opposition to Moslem rule and influence. Hindus and Jains will confide today that it is better to put a calf in a gowshala than have a Moslem eat it.

Cow protection became a political icon for Hindus in their conflicts with Moslems and also when under British rule. Moslems settled in India around the 13th century and can trace their roots to Mogul pastoralists and Arab-Islamic values. Their ritual slaughter of buffalo, sheep and goats is looked down on by Hindus, some castes of which, nonetheless, eat meat. According to Srinivas (1968), the whole Brahminic caste is vegetarian. Of the non-vegetarian castes, fish-eaters look down on those who eat goats and sheep, who in turn look down on eaters of poultry and pigs, who look down on beef-eaters.

Moslems, under British rule, fought successfully to have their religious freedom of ritual slaughter upheld. The British wanted pre-slaughter stunning for humane reasons, but this was not part of sacrificial ritual slaughter under Islamic law. Pre-slaughter stunning eliminates the need to cast the animal onto the ground prior to having its throat cut, thus eliminating much fear associated with being cast.

Native cattle are intelligent and empathetic: Here a cow who cares for calves all day takes them to greet their mothers coming in from grazing in the jungle, and then to be nursed.

For Mohandas Gandhi, cow protection was an important aspect of Indian independence from British colonial rule, figuring in the return to traditional values. He wrote:
"The central fact of Hinduism is cow protection. Cow protection to me is one of the most wonderful phenomenon [sic] in human evolution. It takes the human being beyond his species. The cow to me means the entire subhuman world. Man through the cow is enjoined to realize his identity with all that lives.... Protection of the cow means the protection of the whole dumb creation of God.... Cow protection is the gift of Hinduism to the world. And Hinduism will live as long as there are Hindus to protect the cow. Hindus will be judged not by their *tilaks*, not by the correct chanting of *mantras*, not by their pilgrimages, not by their most punctilious observance of caste rules but by their ability to protect the cow." (Gandhi 1954).

Srinivas (1968) believes that humanitarianism (or what I would call compassion without self-interest) is a Western value not evident in his country because India cannot yet embrace a value embodying concern for all human beings irrespective of caste, religion, age, sex and economic position; or, for all beings irrespective of species, economic, religious or other human-centered value.

Lodrick (1981), in reviewing this history of animal care and shelters in India, concludes that, "Buddhism, although the major vehicle for the spread of the ahimsa concept throughout India and indeed throughout much of Asia, never carried the doctrine to the extremes of Jainism. In Buddhist thinking, ahimsa became a positive adjunct to moral conduct stemming from the cardinal virtue of compassion, rather than the all-encompassing negative principle of non-activity of the Jains."

Humanitarian concerns over animal slaughter and attempts to modernize slaughtering facilities to make them more humane, sanitary, less wasteful and causing less pollution have been opposed by both Moslems and Hindus for religious and political reasons. Moslems see it as threatening their religious freedom (by the adoption of pre-slaughter stunning) and many Hindus see slaughter modernization as a threat to traditional values, totems, taboos, and even national identity and security.

Such opposition is reminiscent of the Hindu cow protection movement that arose in opposition to British rule and the proposed slaughter of cattle as part and parcel of economic development and modernization. Now under the pressures of trade liberalization and an emerging global market economy that is being pushed by the World Trade Organization, efforts to modernize livestock slaughter are being renewed; and opposition intensifies.

The Indian veterinary profession needs to have full government support for developing the livestock and poultry sectors not primarily to produce meat for export and urban consumption, but to integrate humane livestock and poultry husbandry practices with ecologically sound and sustainable, organic (chemical-free) crop and fodder production, and in the process, enable the rural poor to become more self-reliant. It is unwise economically and ecologically, and also socially unjust, to raise any species of farm animal in India (or in any other country for that matter) primarily for meat, eggs or dairy products (Fox 1997). More animal fat and protein for the rich means less bread or grains for the poor. A major goal should be to reduce the overall livestock population to facilitate ecological restoration. Increasing the productivity and health of milk cows and goats through selective breeding and husbandry improvements also needs more concerted and effective attention, and financing. Meat from male offspring and non-productive females ought to be a by-product rather than a primary product, and either be consumed locally or marketed to the meat-consuming sectors. The tempting rationale to raise livestock and poultry for their meat to supply urban markets and for export to gain foreign exchange revenue—a rationale being vigorously promoted by multinational banks and transnational corporations as the way to prosperity for India and other developing countries—must be resisted, because it is not sustainable, even in the developed world.

The flaw in the principle of *ahimsa*, when it takes precedence over compassion, is that it becomes a contradiction. By excluding compassion from ahimsa and refusing to accept humane killing of incurably sick, injured and suffering animals, the principle of ahimsa is violated. The reason for this is purely selfish, i.e., to avoid defiling oneself by defiling the animal in taking its life. This aspect of India's "sacred cow complex" cannot be subject to the light of cool reason and compassion when broached to orthodox Jains and Hindus. After all, it is against the law. Though many will accept that the economic inefficiencies of India's livestock and dairy industries are in large part due to the dilemma as to what to do with millions of nonproductive cattle that compete with productive animals for feed, water, and veterinary care, and are short-changed for economic reasons, the resistance to killing nonproductive cattle who are suffering, or have no feed, results in greater suffering.

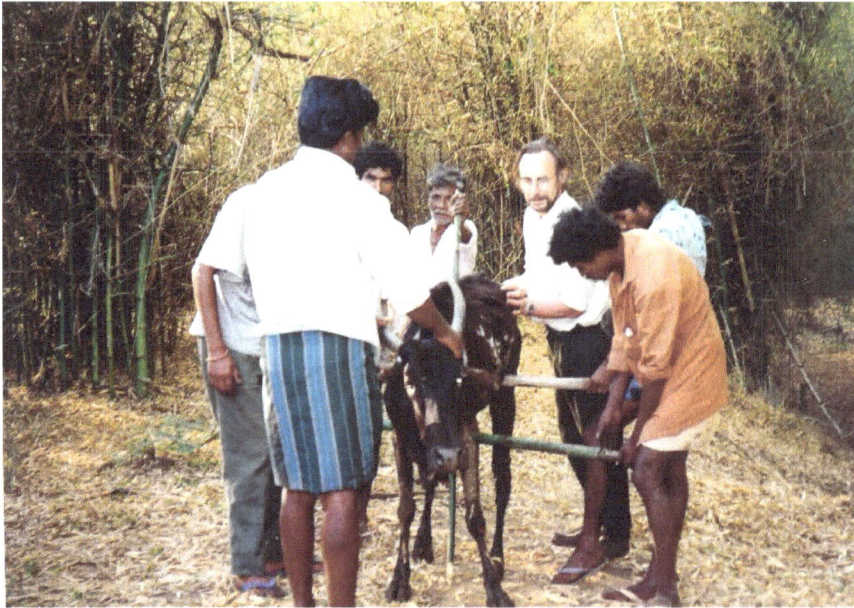

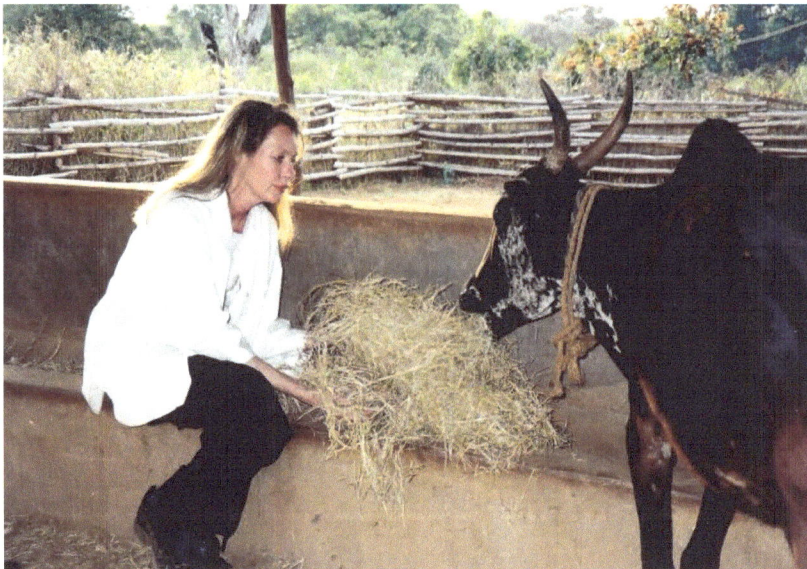

*Author Fox rescues a starving Native 'scrub' cow who had collapsed
in the jungle, and co-author Krantz gives hay to another rescued cow
at her refuge in Tamil Nadu.*

People also tend to confuse *ahimsa* with *aghnya*, the doctrine of non-killing. In the name of compassion, incurably ill and injured animals, those creatures suffering because of old age, and sometimes even those who are newborn, but can not be provided adequate food, should be humanely killed. Compassion must take precedence over both *aghnya* and *ahimsa*, otherwise India will never develop a humane and sustainable agriculture; her sacred cows will continue to suffer until humanity evolves into a more empathic state, or the entire system collapses.

There are ecologically valid and humane reasons for India coming to accept the humane slaughter of cattle as a vital population-control measure, and to see the wisdom of establishing small slaughterhouses in states where cow slaughter is prohibited. There are no simple solutions to the plight of India's cows and their offspring, but with reason and compassion, much suffering could be alleviated.

Agricultural Modernization, Politics and Cattle Welfare

As India shifts to a more capital-intensive industrial agriculture, countless native cows become surplus and urban scavengers for their impoverished owners, and rare breeds become extinct. Many native peoples have been made landless by agricultural "modernization" and migrate in increasing numbers to the cities along with their few animals and possessions. The high cattle population in the nation's capital Delhi is evidence enough. In 1995 some 50 cattle per day were killed or severely injured by traffic (*Kare Newsletter*, 1995).

The Prevention of Cow Slaughter Act of 1955, which allows the slaughter of cattle that are diseased, disabled, or more than 15 years old, allegedly resulted in young, nonproductive cows having their legs hacked and broken so they could be legally slaughtered. The Bharatiya Janata Party (BJP) banned all slaughter of the bovine species when it gained control of Delhi in 1994, purportedly to tighten various laxities in the prohibition of cow slaughter. The BJP voiced Mohandas Gandhi who told all India in 1921 that, "Hindus will be judged ... by their ability to protect the cow."

During the tumultuous 1996 elections, the Vishnu Hindu Parishad (VHP) party, "ignoring the facts and problems" of cattle overpopulation, starvation, disease and suffering, according to *India Today* (January 11, 1996), launched an anti-cattle slaughter campaign. At a rally one sadhu exclaimed, "We shall cut off the heads of those who shed a single drop of cow's blood." Another party leader proclaimed, "The blood of cows has polluted every river."

According to *India Today*, the VHP claims that:

! The trembling and wailing of the cows being slaughtered lead to earthquakes.

! Cow urine can cure cancer, impotence, sexually transmitted diseases, liver problems, tuberculosis, polio, and obesity.

! Eating red meat causes blindness, skin diseases and heart attacks.

! It also results in divorce because eating red meat causes precocious sensuality in children, which later leads to impotence and ultimately divorce.

Opponents believe the VHP/BJP should do something to protect starving cows that wander the streets and get killed and injured by motorists in cities like Delhi where they are in power and remember that beef is an important protein source for the poor. According to a 1992 Indian Market Research Bureau survey reported in this article, 74.2% of urban households are non-vegetarian, the majority consuming mutton, fish, and chicken, and some 12.7% beef (How much is buffalo meat is not clear).

When the BJP won control of the central government in May 1996, the new President Shankar Sharma announced in his opening of Parliament address a total ban nationwide on cow slaughter as one of the new government's policy agendas. One member of the opposing Congress party rose to object, saying such a policy contravened India's secular constitution, which guarantees equal rights to all religions.

India is now at a crossroads where the choice is between rural sustainability and industrial growth and productivity. It is clear which road India is now taking. India exports massive amounts of animal produce—millions of tons of dairy products, hides, bones, horns, hooves, meat, poultry and eggs—even to developed countries like the United States, Australia, and the United Kingdom. The toxic chemicals that most of India's tanneries continue to discharge into rivers and watersheds cause serious ecological and human health problems. While some 200 million people are malnourished in India, the country exported US$625 million worth of wheat and flour, and US$1.3 billion of rice in 1995 (Lappe et al. 1998).

A letter dated June 20, 1994, addressed to us from the Secretary of the Akhil Bharat Krishi-Goseva Sangh Society of Bombay, which claims to be engaged in the preservation and protection of the "cattle wealth" of India, states:

"Our efforts towards preservation of cattle wealth at the political level are not meeting with the desired success in our country in view of the thick skinned bureaucracy and politicians who are hell bent on *destroying the cattle wealth of our nation at the behest of the meat lobby*, which finds enormous wealth in this activity as also at the behest of FAO, an organ of United Nations which dictates policies in third world countries, *aiming at total destruction of the cattle resources of third world countries*.

"However there is a silver lining to this otherwise discouraging scenario and that silver lining is in the form of our judiciary. Some time back a case instituted in a court in New Delhi involving shifting of a slaughterhouse from one area of Delhi City to another area, the Learned Judge who delivered a judgment in this case has made an excellent analysis of the whole issue and established the legal rights of animals as well as the need for conserving animals for conservation of environment. He has established that the human race, the environment and the animals are interrelated and extinction of animals will spell doom for environment and mankind."

Contrary to this Learned Judge's views on environmental conservation, an almost insoluble problem has been created by the ecological damage caused by over-gazing of cattle, buffalo, sheep, and goats and their diseases and hunger, problems compounded by a lack of fodder and vital grazing land that has been taken over to grow feed and fodder for intensive modernized dairies, buffalo calf meat production and egg and poultry factories, and for cash-crops. The root of the problem is ideological, and the ideological conflicts between the reasonable and the less reasonable must be resolved. India's "cattle wealth" is first and foremost a family and community matter. The above Delhi judgment is based more on historical tradition than on reality. The expansion of the domestic animal and human populations in India will spell doom if they are not controlled. The monopolistic capitalization of India's "cattle wealth" by developing export markets that are not based on humane, sustainable and socially just methods of animal and plant production (Fox 1997) is unwise and bioethically unacceptable.

What is called for is a *unified sensibility* that integrates the symbolic, material/economic, emotional, social and spiritual components of the human-cow/cattle relationship into a *mutually enhancing symbiosis*. The human side of the relationship is more balanced and equitable when the rights, interests, and welfare of animals are given equal and fair consideration. The ethical inconsistencies in the religious and secular communities' attitudes toward and treatment of cattle and other animals is more evident in

India than in other countries precisely because India is the birthplace of the highest spiritual principles pertaining to animal welfare and yet they are not always put into practice, creating an essentially schizoid situation between the ideal and the real.

In the stifling traffic of New Delhi, a worn out and emaciated bullock pulls an overloaded cart.

Caring for animals and caring for people, for the poor and the hungry, go hand in hand as part of the humane agenda of any democratic society. While this article focuses particularly on India's cattle, the plight of these creatures mirrors the plight of the poor. There are no miracle remedies for hunger and poverty from advances in technology, science, or medicine. The miracle will come not via genetic engineering of animals and plants but through the transformation of humanity into a compassionate, empathic, and responsible life form. A mutually enhancing symbiosis with the Earth community of plants and animals, both wild and domesticated, is our only viable future. Our hope lies in our capacity to reconnect empathically with all living beings and to use sound science and policies as our instruments, and compassion as our compass.

Indian artist Amit Ambala's pertinent depiction of the human-cow relationship.

Addenda

India, Source Of Mad Cow Disease?

A highly controversial, but plausible hypothesis that the mad cow disease epidemic that devastated the UK's cattle industry, harmed many other countries and infected people with this fatal brain disease, actually originated in India was published by Drs. Alan and Nancy Colchester in the British medical journal *Lancet* (2005, vol. 366, Pp. 856-861). Since cases of C-J disease have been reported in people in India, the exposure of cattle to potentially infective human remains, especially along the banks of the polluted Ganges river, is a high probability, especially considering the co-mingling of cattle and people everywhere. Often poorly incinerated human remains from funeral pyres are scattered over the Ganges river, and the entire, un-cremated bodies of children, saints, religious leaders, and people who have died from various diseases, weighted down with stones or banana trees, are also thrown into the water (see K. Tillotson, startribune.com/world, Jan. 16, 2006, pA9 and A13). Frequent exposure to, and ingestion of, human remains increases the statistical probability of the emergence of a zoonotic disease, as is proposed in the case of C-J variant disease in humans of bovine origin. It would take but one cow to develop this disease from the remains of an infected human, and for the remains of that cow to be processed and exported to the UK or any other country to start an epidemic, because the agent responsible for this disease, a prion, is resistant to extremely high temperature processing and desiccation. Bone meal and other cattle parts, such as horns, hides, hooves, and blood, are major Indian export commodities (along with dairy products and now increasingly popular prepared foods and sauces for human consumption), the bone meal being used in livestock feed and as a human dietary supplement or food additive.

India Becoming World's Leading Beef Exporter

In October 2011 the USDA forecast that Australia and Brazil would remain the largest exporters of beef in 2012 with exports of 1.38 million tonnes each, followed by India at 1.28 million tonnes and the US with 1.25 mt. But now, according to the U.S. Department of Agriculture's international marketing forecasts and India's own *Financial Express* (April 8, 2012), India is poised to become the world's largest beef exporter by 2013. Factors other than price would also help to expand Indian buffalo meat exports, particularly to North Africa and the Middle East.

Meat is slaughtered following Halal standards and the lean character of buffalo meat has several positive blending characteristics sought by processors, the USDA reported. But when I observed buffalo slaughter in a large facility in India these standards were in question and were pointed out to me by the concerned Indian veterinarian who accompanied me. The prone animals were having their throats hacked while they struggled.

India has a herd of 185 million buffalo. Ample supplies and relative weak domestic demand mean that it relies on export markets to absorb increased production for the valued buffalo milk.

The rising demand for low-cost Indian buffalo meat by 'price sensitive' importers such as Vietnam, Africa, Middle East, and Southeast Asia resulted in some 1.52 million tones of buffalo meat being exported in 2012. Further market expansion may be limited by trade restrictions being placed on India because of Foot & Mouth Disease. The export of meat and other foods from India where millions are malnourished is an inescapable irony. Raising buffalo calves for this new export market may soon be challenged by an informed and concerned public, raising questions about how these surplus buffalo calves can ever be fed and watered sustainably in a country where land, feed and water are in ever shorter supply. Also what safeguards if any are in place to prevent the mingling of this meat from buffaloes with the meat from illegally slaughtered calves and bullocks from dairy cows? The buffalo beef market is another tempting opportunity for the cattle mafia of India.

According to a 2013 United Nations Children's Fund report, some 61.7 million, or 48% of all children in India, are stunted physically and immunologically and mentally impaired because of pre-and postnatal malnutrition. It is a tragic irony that India is now leads the U.S. and Argentina as the world's leading exporter of beef (from buffalo).

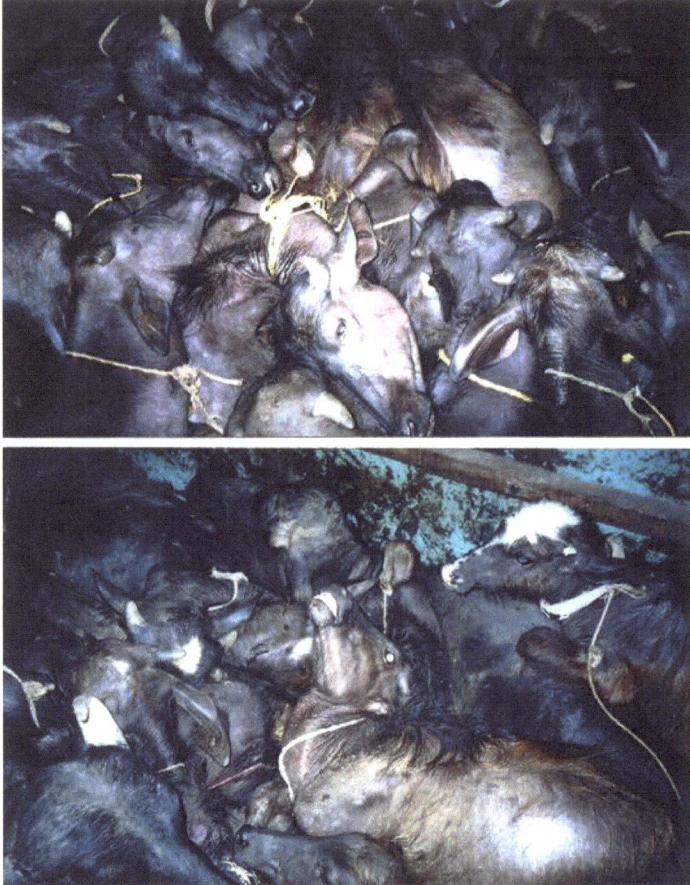

Cruel packing of buffalo and cattle in truck destined for the slaughter house and export beef market.

India's Misuse of Antibiotics a Threat to the World

In 2010 epidemiologists recognized a new strain of "superbugs" carrying a genetic code first identified in India — NDM1 (or New Delhi metallo-beta lactamase 1), which has since spread around the world to countries including France, Japan, Oman, and the United States.

In his *New York Times* (Dec. 3,2014) article "Superbugs Kill India's Babies and Pose an Overseas Threat," Gardiner Harris writes: "A growing chorus of researchers say the evidence is now overwhelming that a significant share of the bacteria present in India — in its water, sewage, animals, soil and even its mothers — are immune to nearly all antibiotics." Furthermore, "India's dreadful sanitation, uncontrolled use of antibiotics and overcrowding, coupled with a complete lack of monitoring the problem, has created a tsunami of antibiotic resistance that is reaching just about every country in the world," said Dr. Timothy R. Walsh, a professor of microbiology at Cardiff University. "Bacteria spread easily in India, experts say, because half of Indians defecate outdoors, and much of the sewage generated by those who do use toilets is untreated. As a result, Indians have among the highest rates of bacterial infections in the world and collectively take more antibiotics, which are sold over the counter here, than any other nationality. Just as worrisome has been the rapid growth of India's industrialized animal husbandry, where antibiotics are widespread. Most large chicken farms here use feed laced with antibiotics banned for use in animals in the United States. A New Delhi science group recently found antibiotic residues in 40 percent of chicken samples tested."

The overuse, improper use and inadequate oversight of antibiotic use in the livestock sector has been a problem for decades In India, a major manufacturer and exporter of such drugs. Other drugs used in the livestock sector are also of concern notably the near extermination across the country of the vulture, poisoned by diclofenac which is injected widely into cattle to treat lameness and other inflammatory and often misdiagnosed conditions and which remains in the animals' body parts consumed at outdoor dump sites by these birds. Another concern is the injection of oxytocin into dairy cows to stimulate let-down of milk which can cause violent smooth muscle contractions and prolapse of the vagina and uterus, and may put consumers at risk.

References

Crolty, R. (1980) *Cattle, Economics and Development.* New York: Oxford University Press.

Fox, M.W. (1996) *The Boundless Circle: Caring for Creatures and Creation.* Wheaton, IL: Quest Books.

Fox, M.W. (1997) *Eating With Conscience: The Bioethics of Food.* Troutdale, OR: New Sage Press. See

Fox, M.W. (2001) *Bringing Life to Ethics: Global Bioethics for a Humane Society.* Albany, NY: State University of New York Press.

Gandhi, M.K.(1954) *How to Serve the Cow.* Ahmedabad: Navajivan Publishing House, pp.3-4.

Harris, M.(1991) *Cannibals and Kings: The Origins of Cultures.* New York: Vintage Books.

Kindness to Animals and Respect for the Environment (KARE) (1995) *Expose Newsletter.* New Delhi,4/1, July.

Lappe, F.M, Collins, J., and Rosset, P. (1998) *World Hunger: Twelve Myths.* Second edition, New York: Grove Press.

Lodrick, D. (1981) *Sacred Cows, Sacred Places. Origins and Survivals of Animal Homes in India.* Berkley, CA: University of California Press.

Report of the Expert Committee on Development of the Meat Industry.(1987) New Delhi, Ministry of Agriculture and Co-operation

Shephard, P. (1996) *The Others: How Animals Make Us Human.* New York, Island Press.

Simoons, F. J. (1961) *Eat Not This Flesh.* Madison, WI: University of Wisconsin Press.

Srinivas, M. (1968) *Social Changes in Modern India.* Los Angeles: University of California Press.

Chapter 4
Elephant Truths: The Saga Of Loki & Captured Indian Elephants

IPAN

To Help One Elephant

Captured Loki, a gentle animal.

Deanna Krantz, as director of India Project for Animals and Nature (IPAN), first became involved in Asian elephant problems in the Nilgiris, South India, in 1998 after she was called in by concerned local people and some Forest Department staff over the rapidly deteriorating condition of a recently captured makhna (a rare tuskless bull elephant), aged around 35 years old. Deanna and her staff called this docile makhna, who was in urgent need of expert veterinary care and proper nutrition, 'Loki.'

Loki in his small enclosure, unable to walk or lie down, with severe injuries to all four legs and infected gore-wound on his shoulder, shortly after capture. One severed tendon protrudes from his ankle. White antibiotic powder covers his injuries.

Pus and blood draining down Loki's chain-gouged fore-legs, 2 months after capture.

Severed tendon on Loki's left fore-leg, 2 months after capture.

The saga of this elephant reached the Indian Supreme Court, and tribal staff sang out his name into the jungle night as they gathered around a ritual fire to pray for his freedom. The Forest Department insisted that Loki's capture was necessary because he was a "rogue." The Forest Department alleged that this makhna had killed up to 35 people, raided crops, and wrecked tea plantations. Some three years after his cruel and injurious capture by the State Forest Department of Tamil Nadu (Deanna secured a video film copy of Loki's capture taken by a local TV station that was hired by the Forest Department to document this event), it was reported, contrary to the reasons why he was caught, that a rich timber merchant had bribed the Forest Department to catch him. An article in the Tamil Newspaper, entitled "Makhna Would Not Have Been Captured, Environmentalists Feel" (*Dhina Malar*, January 12, 2002.) stated "Forest officials have found 100,000 acres of natural forest land has been encroached and trees felled, like rosewood and various fruit-bearing trees that supported colonies of Bonnet macaque and Langur monkeys and other wildlife. Environmentalists feel that if the makhna had not been captured, this forest destruction would not have happened because tree cutters feared his presence. It was reported that "Southern Wood Industries owner Mr. Gudalur spent a few 'laks' to persuade officials to capture the makhna."

That Loki the makhna was removed from his forest so that it could be cut down is indeed a tragedy. Loki's saga began with many headlines. One of the most factual and objective articles was published in *The Hindu*, March 8, 1999, by S. Bharath Kumar entitled "Global plea to free confined elephant," in which Kumar wrote:

"Pressure from abroad is mounting for the release of Loki, the 'makhna' (tuskless) elephant, from the Mudumalai elephant camp, where it has been reportedly 'ill-treated' by the authorities of the Tamil Nadu Forest Department since his capture over six months ago.

Nearly 60 (sic) #* U.S. Congressmen have in two appeals pleaded that the elephant be given proper care and attention. The U.S. House of Representatives has said: 'Beatings, neglect and confinement to a kraal in which the elephant is unable to walk has resulted in one of the worst cases of animal cruelty ever documented.

The 'makhna' is extremely malnourished and it is critical that he be released immediately so that his wounds are treated. A suitable sanctuary and support is available for the makhna upon his release.

These appeals were faxed to the Indian Ambassador to the U.S., Mr. Naresh Chandra, and copies sent to the Union Forest Minister, Mr. Suresh Prabhu, and his Tamil Nadu counterpart, Mr. Palaniswamy. Sources in the Tamil Nadu Forest Department confirmed the receipt of the message and said there was
'tremendous pressure' from certain quarters not to release the elephant.

It was in July that the pachyderm roaming in the Tamil Nadu forests was captured as it had become a 'rogue.' It was then shackled and dragged across 40 km to the Mudumalai camp over a period of eight days, all along gored by 'kumkis.'** So bad were the injuries it suffered that since then it has not sat and slept, according to an official who did not want to be identified. The 'makhna,' like other elephants in the camp, is malnourished. Says a mahout, with tears in his eyes, that officials manning the camp coolly line their pockets with the funds meant to buy the elephant food.

An American, Ms. Deanna Krantz, who is an animal activist and runs a refuge for cattle and dogs, wanted to provide food for the 'makhna,' but she was prevented from doing so by the officials."

* Actually 30 Congressional representatives and one Senator signed the petition.

** Kumkis are large-tusked, adult bull elephants who are trained to be the camp enforcers when any elephant tries to break free, revolts, or is a mother who will not surrender her captive-born calf to be broken into service.

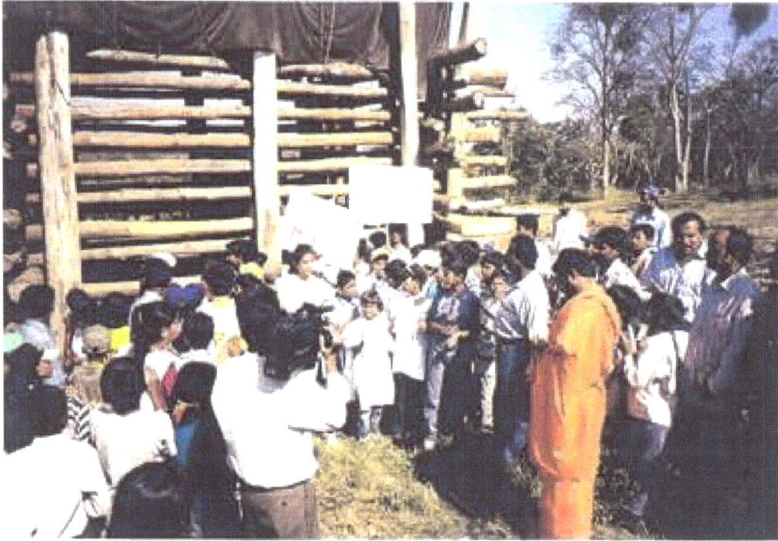

School children demonstrating before Loki's kraal as Jain Swami talks about compassion.

Schoolchildren form a circle around Loki's kraal.

The IPAN team that initially treated Loki intuitively knew that he was never a killer. He was an extremely gentle and compliant animal. He would trumpet a loud greeting whenever he heard IPAN's jeep coming with food and staff to treat his wounds.

As Loki's mobility and strength improved as a result of IPAN's daily care under the direction of Krantz, an irrevocable decision was made by the Tamil Nadu State Forest Department authorities. The mahouts were ordered to beat him daily until he obeyed the command to lie down, the first step of elephant "training." This training was begun following approval by the Forest Department's consultation with India's most famous and internationally recognized Brahmin of elephant veterinarians, the now late Dr. Krishnamurthy, who was featured in the PBS documentary *The Elephant Men*. IPAN took video and other documentation of his non-sterile and inappropriate veterinary procedures, and evidence of other instances of cruelty, neglect and lack of veterinary care involving several other elephants at this camp, including the blinding of one after being chained and beaten, and another dying from a massive tapeworm infestation.

IPAN volunteer veterinarian Dr. James Mahoney, who had been attending to Loki every day, arrived one morning to give treatment and found two mahouts beating the elephant on his leg wounds, yelling at Loki (whom they named 'Murthy' in honor of Dr. Krishnamurthy) to "bite" (lie down). This systematic cruelty, which Dr. Mahoney recorded on audio-tape, along with the elephant's cries, lasted for 45 minutes. A horizontal, partially healed cut across the base of his trunk was evidence that he had been hacked there with a machete, another "standard procedure" of elephant "training." Loki never could lie down since the tendons in his legs were so damaged by chains, and never once did he ever lie down and rest during the entire seven months that the was confined in the small kraal at this elephant camp.

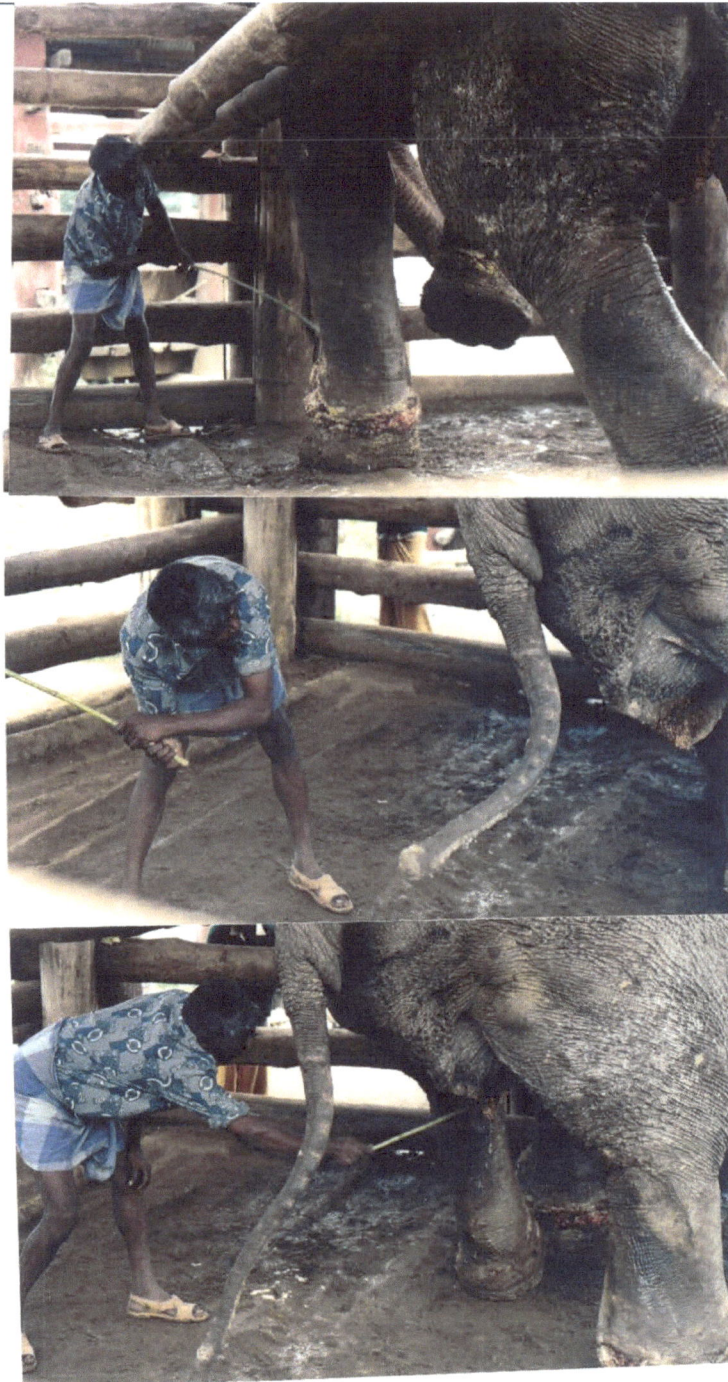

Loki receiving systematic daily beatings on his injured legs ordered to make him compliant and to lie down on command, which he was physically incapable of doing.

Soon after this beating incident, on Christmas Day 1998, Dr. Mahoney, Krantz and her staff—who had brought fresh green food for months for the makhna that they collected every day from the fields, often in monsoon rains, at 5:30 am in the morning—were prohibited by the Forest Department from further contact with the makhna.

Thanks to the support of Union Government Minister Maneka Gandhi (who had initially sided with the authorities who spread disinformation about IPAN), and Attorney G. Ragendran, IPAN won a High Court ruling after a hiatus of three years to be allowed back into the elephant camp and see to Loki's condition. In early January 2002, IPAN's team, headed by veterinarian Dr. James Mahoney and internationally recognized elephant care specialist Alan Roocroft with the San Diego Zoological Society evaluated Loki's condition. They began a course of treatment on Loki's crippled and chronically infected legs and feet, made recommendations to improve his overall well-being—chains tied to his legs were causing further lesions—and leveling out his resting area which was on a slope, a very uncomfortable spot for an animal weighing several tons who was also a permanent cripple. The Wildlife Warden told Mr. Roocroft that Loki's resting area would be seen to immediately. But it never was done.

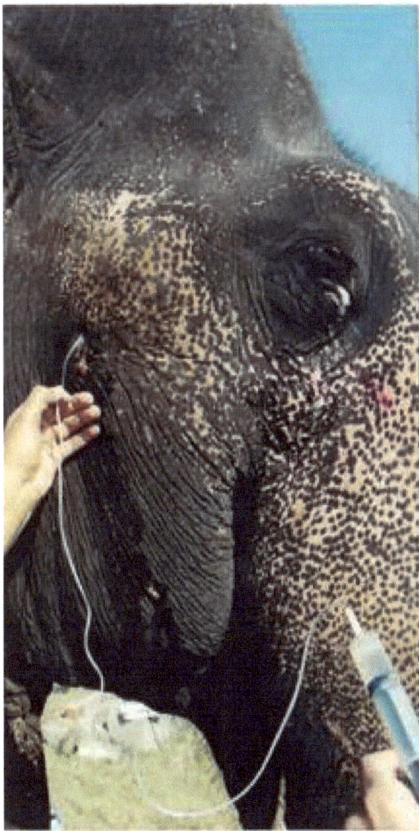

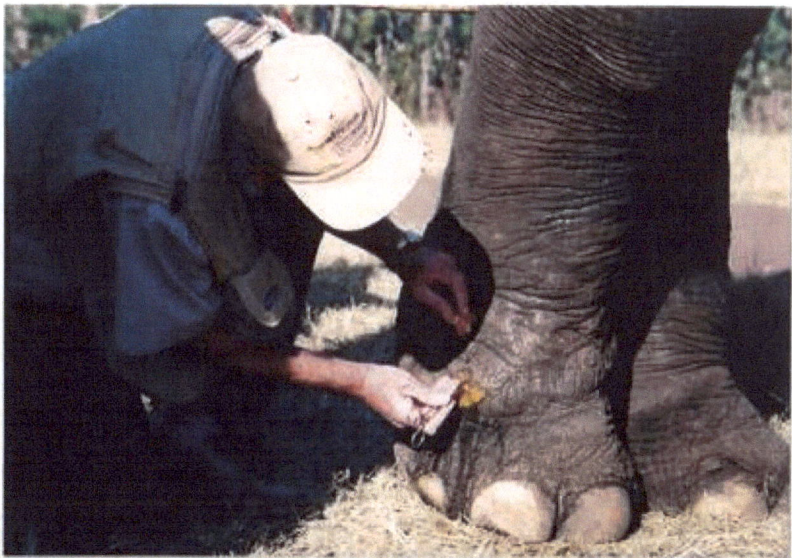

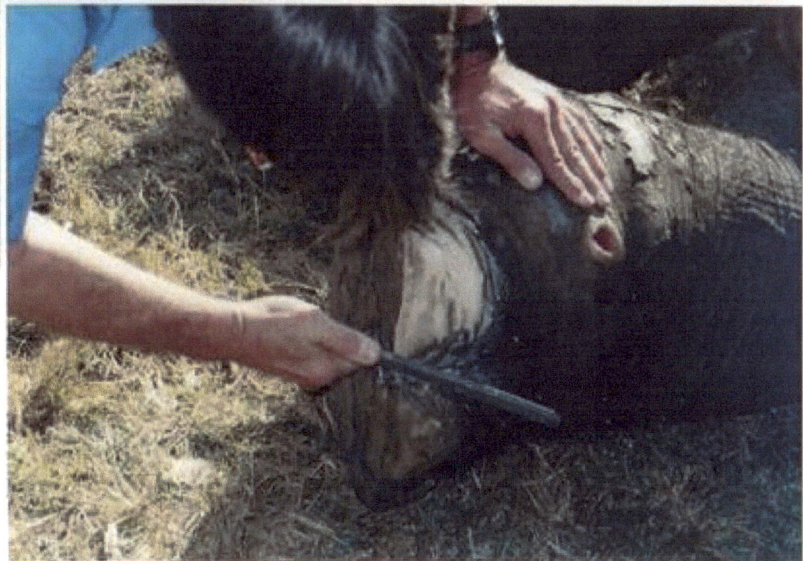

ı hiatus of 3 years, a High Court ruling allows
ıhoney to evaluate Loki's condition and treat
ıunds, deep abscesses, and a new deeply infected
l on his face. Elephant expert Alan Roocroft
his feet

After a hiatus of 3 years, a High Court Ruling allowed Dr. Mahoney to evaluate Loki's condition and treat his old wounds, deep abscesses and a new infected wound on his face. Dr. Roocroft treated his rotting nails and feet.

After several days' intensive treatment, the State Forest Department, rather than collaborating, chose to file a court injunction barring the IPAN team from further contact with the elephant. The Supreme Court gave the Forest Department 15 days to evaluate and report on the makhna's condition. The official report submitted to the Court, according to the *Dhina Malar* newspaper February 20, 2002, would include the fact that the three veterinarians, whom the Forest Department had brought in, had "completely healed the elephant with a new mixture of medicines containing olive oil, honey and barley."

Loki being treated.

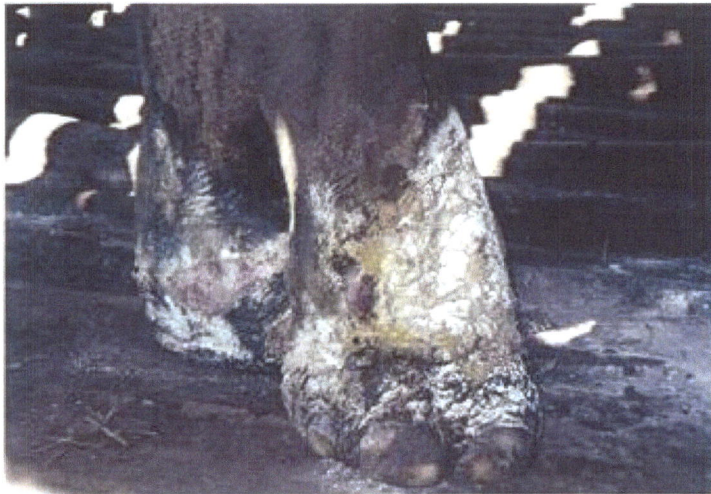

Unhealed wound and swollen right leg 1 week before release in chains.

Loki's wounds on hind legs from chains, partially healed after 4 months of massive infection.

Other Elephant Camp Cruelties

No one any distance from a small clearing in the Indian jungle can hear the cries of elephants. But, the world must hear. Stretched out in chains, elephants are routinely beaten at this and other camps, sometimes by a gang of 6-12 men for up to an hour, by which time the men, some drunk, are exhausted. So elephants, wild caught and captive born, are still being broken and disciplined for being disobedient.

Many visitors witnessed a fiasco at this camp where a mother elephant was being prematurely separated from her 21-month old calf. The mother was semi-tranquilized and was tied with ropes to two trained bull elephants, but she struggled and protested as her infant was forced away by other elephants and men. The mother screamed when her infant screamed, held captive in a small log enclosure where he would be broken in spirit to depend on man, and obey.

After chasing and beating the mother and controlling her with Kumkhis (above) her infant was eventually separated from her, kept in chains and subjected to dominance training with a whipping cane.

The Kumkhis (bull elephants), naturally protective of their own kind, became highly agitated as the mother, screaming and fighting to be free and with her baby, was beaten with sticks from face to flank until she collapsed. One may wonder which would be the next Kumkhi to go berserk, or swipe out with his trunk, or swing a tusk at the elephant tormentors, and be chained, starved and beaten.

The widely revered elephant god of India is called Ganesh. Road-side shrines to this animistic deity of Hindu religious tradition can be found throughout the country, as exemplified by the decorated statue on a quiet country road depicted below.

But a real life elephant called Ganesh, who had worked many years at this camp, swung at a visiting official—the Chief Wildlife Warden—and broke the man's tooth. For such "rogue" behavior, and for shaming his mahout handler and camp, a dozen mahouts chained Ganesh to a tree and beat him for almost an hour, to "take out his nuts and bolts," they said in their Tamil tongue. They blinded him in one eye. Subsequently Ganesh was kept chained in one place and fed starvation rations for no less than five months. This is standard, traditional treatment for disobedience and injurious actions and intentions. Ganesh turned into a living skeleton, like Subramanian who was also blinded in one eye, and Vigiai, and other elephants before him.

In February 2001, I met Veterinarian Dr. Dennis Schmidt, who is associated with Barnum & Bailey and Ringling Brothers Circus, near this elephant camp after he had been there with another veterinarian from Germany, a Dr. Thomas Hildebrand from Berlin's Institute of Zoological and Wildlife Research. Elephant management consultant Heidi Riddle from the US was with them. They told me initially that they had been "conducting a workshop" at this elephant camp. Ms. Riddle, with husband Scott Riddle, run Riddle's Elephant and Wildlife Sanctuary in Arkansas and the "International School for Elephant Management" and is a past President of the US Elephant Managers Association (EMA). The EMA has close ties with the American Zoo Association. Scott Riddle had been forced to leave the UK Blackpool Zoo for using and allegedly promoting cruel elephant training and handling methods.

After Ms. Riddle told me they had been doing ultrasound tests and giving an elephant management training course at the Theppakadu elephant camp, the vocal Dr. Hildebrand told me that they had been collecting semen from the elephant bulls at this camp by putting one arm up their rectums trying to stimulate ejaculation while the animals were tightly shackled and beaten if they moved. The German veterinarian was disappointed with what little semen they were able to collect. When I pointed out that the bull elephants were chronically malnourished, emaciated and kept chained to a tree for 17-18 hours a day, so they had no libido, Ms. Riddle said elephants in the wild are also often thin and quickly defended this elephant camp. But it is only nursing mothers and very old elephants in the wild who sometimes become naturally thin.

Ms. Riddle and Dr. Schmidt dismissed my concerns about the mistreatment of the elephants at this camp, contending that all was well and questioned my knowledge of elephants. They also argued that what India Project for Animals and Nature (IPAN) had documented concerning the severe injuries sustained to the wild Makhna, in the course of capturing him, and his subsequent inadequate veterinary care, confinement for 7 months in a 16x16-foot log stockade, and repeated beatings, was all overblown and sensationalistic, because IPAN had no expertise in elephant management and care. Yet it was the IPAN team that worked long hours for several months providing Loki with food, medicines and veterinary care, and essentially saved his life.

To exist but not to live: Bull elephants in Sate camps are hobbled and tethered on short chains most of their days and nights.

Veterinarian Dr. M. Sugumaran, whose medical and surgical work for the benefit of the animals and the rural poor in the Nilgiris we continue to fund, sent the following report:

Elephants at Teppakkadu camp: Report from IPAN's veterinarian, July 2011

There are 27 captive elephants being maintained at Teppakkadu elephant camp. For good grazing and easy management they are divided in to two or three groups. A few elephants, including Loki, are maintained at Abayarayam camp around five kilometers away from Teppakkadu, and few are taken to interior camps such as Game hut and Bombox for grazing and exercise.

There are two infants, one is a five year old and the other is one year old; both are cared for at Teppakkadu camp and the five year old is under training. Both were rescued from the forest. As per my knowledge and available information the captive elephants are not being bred during the last few years.

Loki looks healthy and I am told he often evades tethering and disappears in the jungle and is brought back to the camp by engaging kumkis on several occasions.

I agree with your last statement that Loki is not a happy captive elephant that is why he evades tethering.

Loki pulling a long, heavy drag-chain at the elephant camp where some chained elephants have wandered into the surrounding jungle and were not recovered, being found starved to death with their chains caught around a tree.

At Teppakadu elephant camp a rescue and treatment facility for wild animals is being commenced by the Forest department last week. A vehicle for rescuing small wild animals is also available. A reasonable progress is happening but they are having much reservation on people like us because they do not want transparency. Anyhow I am keeping my touch on as much as possible. In my opinion your presence and initiation directly and indirectly is the base for these changes. I am sure that but for your initiation nothing could have moved for the cause of wild animal welfare.

In the wild one young tusker was poached for tusks last week, with that the toll for tusk goes up to three for this year but the department hesitates to disclose openly.

Man and elephant conflicts arise on a day to day basis. Last month two persons were killed and two were injured by wild elephants in the Gudalur area.

Two wild elephants rampaged in Mysore city last month and killed one person.

Yesterday (6 July 2011) wild elephants damaged two houses at two different incidents near Gudalur. As you always mention that the fragmentation of herds due to lack of food in the wild and interference in their corridor passage is the major cause for these types of problems.

IPAN's in-field veterinarian made this second report on July 13, 2011:

Visited the elephant camp on 12 July evening along with 19 student of Madras veterinary college. We were able to see eight adult elephants and the two calves who were rescued from the jungle.

We also saw our Loki at Teppakkadu camp. As one of his legs (left hind) had swelling, he has been brought for treatment from the other camp.

As I mentioned last week the five year old female calf is under training. I am told that she is being treated well and the other infant is under care in a separate well ventilated room, one tribal family being entrusted with that work.

Recent events:

The forest department people (forest staffs and veterinarians with some elephant researchers from the World Wildlife Fund have been applying radio collars to wild elephants to monitor their movements. While fitting a radio collar onto a tranquillized elephant, another elephant died from the injuries sustained due to a fall after tranquillization. This work was being carried out in the forest near Coimbatore north east to Mudumalai.

Two persons were killed by elephants on Monday (11 July 2011) in separate incidents, one by a wild elephants near Gudalur and the other person was killed by a trained elephant in a camp near Coimbatore.

Good news:

Madras High Court issued orders to the Government to take action to vacate the resorts and other occupations from Bokkapuram area, Mavanallah area and Vazhaithottam area to clear the elephant passages (corridors) and every one is eagerly waiting for proper action without affecting the tribal people and poor farmers.

For details, please browse www.forests.tn.nic.in.

The Realities of Elephant Care and Conservation

Cultures resistant to reason and sound science often adhere to superstitious and outmoded beliefs, and magical thinking. The greater the adherence and resistance, the greater is the tension and dissonance between ritual and belief, and reality and suffering, as when images of elephant gods are revered, but wild elephants are being killed, and worse, captured and enslaved to a degraded existence in a zoo, circus, or in chains to carry tourists and the last logs out of the dying forests. To be enslaved to destroy their own domain is the ultimate injustice and degradation of these pachyderms, who are more ancient, if not more enduring, then we.

Saving the last of the Asian elephants through artificial insemination and in-vitro fertilization, the latest cause of the zoo business, is preservation, not conservation. These great souls are becoming extinct, including one of India's largest wild population. They so rarely breed and even more rarely produce viable offspring when held in the typical urban zoo environment, in work camps for timber and tourism, and in traveling circuses.

The Fate of the Wild and the Captive

Thanks to IPAN's network of in-field tribal and Forest Department informants, we have collected video and photo-documentation of elephant ivory poaching, electrocution, and killing with homemade country guns and bombs for crop-raiding; also illegal tree-cutting, river diversion for irrigation, fencing, land development and sand and rock quarrying within the elephants' ostensibly protected domain. We estimate that over 60 elephants had been killed over an 18 month period in our vicinity between 2001-2002. Carcasses of killed elephants and other wildlife killed by poison bait, like hyenas, are often burned by Forest Department field staff to hide the evidence for fear of demotion and loss of pay for not apprehending the killers.

One rich land owner who was involved in the killing of an elephant by electrocution on his property had his prison sentence and charge-sheet expunged by an attorney, who heads the "Save the Nilgiris" conservation society and has served as India's leading legal representative with the International Union for the Conservation of Nature.

Recent National Geographic reports and PBS TV documentaries on the elephants and this bioregion, and of those involved in their captive care and conservation in the wild, are disturbingly sanguine in their appraisal of elephant welfare and conservation.

The truth about the plight of elephants, captive and wild, is distorted by so many converging and often conflicting interests, as well as vested interests, from the "sacred" to the secular and the highly commercial. Tribal elders have told me that they remember a time when they lived in harmony with the elephants. There was never such a thing as a "rogue" elephant. In but one elephant generation they have become crop raiders and people killers. Such reactions in a highly evolved mammal are to be expected considering how their habitat has been fragmented and depleted, and how often they have witnessed their herd-mates and sons and fathers being shot for their ivory, often defending them from poachers where they fell.

We barely comprehend the dimensions of elephant consciousness, of elephant sensory, cognitive, social and emotional realities; "bush wisdom," communication and also what vital role they play in the ecosystems that they have helped regenerate for millennia. They are a 'flagship' species, literally King of the jungle. It is a moral imperative for the world community to be involved collaboratively in elephant care and conservation, and not to be shut out either by governmental or nongovernmental organizations when elephants continue to suffer in captivity and their wholesale slaughter in the wild continues.

The dead elephants that IPAN has seen, seem to be saying that extinction—and maybe death by fighting humans, like the mother elephant and two other females who kept poachers from taking the small tusks from her young son during my last week in the Nilgiri Global Biosphere reserve in February 2002— is better than life in a test tube, a zoo, or the Last Circus on Earth. If they cannot exist for themselves, be themselves, and live free in the living jungle, then surely all else is human hocus pocus.

I have felt the pain of others who witnessed an elephant "skidding in her blood" as she tried to run in terror after her mouth had been blown apart by a homemade bomb placed inside a large Jackfruit set out to kill crop-raiding wildlife. I have seen the devotion of Deanna Krantz's staff whose round-the-clock efforts to save an injured, orphaned baby elephant was abruptly terminated by one Forest Department veterinarian, who claimed jurisdiction over this endangered species and contended he was better able to care for the baby. Three days later, she died in the arms of her staff, who had been called back to save her from three terrifying days without appropriate care and love.

Baby elephants, like any human or other mammalian offspring, cannot survive without love, the bonding that comes with tender loving care. Precisely because they are highly immature, slow maturing and precociously empathic, they are highly vulnerable souls. Baby elephants rarely survive in captivity, East and West. The cost of producing one through ART (artificial reproductive technology) and raising the elephant to reproductive maturity, like the baby elephant produced by artificial insemination at the National Zoo, Washington, DC, I would estimate to be around $3 million. Make that $36 million to create a small herd in captivity, for some indeterminate time. Half that amount, if properly used, could save the Nilgiri's entire elephant population—India's largest.

But regrettably those purportedly involved in elephant conservation are aligned with ideologies and vested interests that are as anathema to conservation and environmental solutions as the mafia is to social justice and law and order. The alignment of Western NGOs and governments with institutions and individuals in India who have a monopolistic control of elephant conservation moneys and research and initiatives, is a matter of public record. This monopoly has been called the "elephant mafia" by local Indian observers who see no evident reduction in elephant killing and habitat encroachment and destruction. The Asian elephant situation is worsening in spite of 30 years of research studies and conservation projects, mostly funded by the West. One of Deanna Krantz's staff, Kurumba tribal Madiga, recounted a time in his youth when he knew over 100 adult bull elephants—tuskers—by name. Now, 30 years later, there are few if any adult bulls left in this region in the Nilgiris. It is they who discipline unruly young males who may injure and even kill young female elephants, and get injured or killed themselves from electrocution or falling into trenches dug deep to protect crops.

One well-known Indian Asian elephant biologist has actually built a bungalow in his large, fenced-in estate, located in the center of one major elephant corridor in the Nilgiris. Fittingly, the bungalow is made from stones taken from blasted rock from the construction of a hydroelectric power facility that was opposed internationally by environmentalists because of its harmful environmental impact on elephants and other wildlife, and their habitat. In 2009 this award-winning elephant 'conservationist' Raman Sukumar, from The Indian Institute of Science, Bangalore, sided with a group of scientists who planned to construct an underground neutrino observatory in the heart of this wildlife preserve, which would mean constant traffic with tucks removing tones of rocks for many months, and then an increase in human settlement. We were glad to be part of a coalition that successfully blocked this multi-million-dollar project that had no place in the U.N. designated Nilgiri Global Biosphere Reserve. It is no coincidence that Sukumar had sided with the Tamil Nadu Forest Department during the Loki saga, telling the press that Loki's injuries were self-inflicted because he fought the chains around his ankles, and the alleged Kumkhi gore wounds on his body were in fact old gun-shot wounds. In order to secure the necessary permits for his Institute to continue elephant and other wildlife research in the Nilgiris, and to bring in college students from the US and groups of eco-tourists from the US and Europe that are lucrative enterprises, he had a vested interest in defending the Forest Department and the status quo at the elephant camp. On at least one occasion his research team killed a wild elephant with an overdose of tranquillizer from a dart gun, shooting the wrong animal by mistake. They were to fit a radio collar for tracking in order to study elephant feeding patterns, and covered up the catastrophe by reporting that the elephant fell and impaled herself on a cut tree trunk, according to our tribal informant observers.

Conservation Initiatives

From our investigations, this "elephant mafia" (or establishment) is not so much an organized crime network that profits illegally from the Asian elephant, but a nexus of vested interests and coincidental associations that seem to profit in more ways from the elephants than the elephants evidently benefit from them, in terms of fewer mortalities in the wild and improved welfare, husbandry standards and mahout training and support for those in captivity.

Challenging and Changing the Status Quo

This is probably one of the reasons why other researchers and involved persons have remained silent on various conservation-related issues in the area, a situation we term the "politics of extinction," like some state veterinarians falsifying the cause of elephant deaths in collusion with certain Forest Department officials who have been bribed by rich land owners. Rather than hear what is really going on where scientists rarely tread either in the field (jungle) and talk to the tribals, or at Loki's elephant camp and talk to those mahouts who are not afraid to speak out, the Asian elephant preservation and propagation establishment continues to protect not the elephants, but the status quo; and good money continues to go after bad.

To touch any group or individual in this network as a concerned outsider will meet resistance if there is no money forthcoming, and will face volumes of glossy and self-lauding and self-promoting booklets, reports and magazines on what progress is being made to conserve elephants and other endangered species. It seems like a charade to preserve not elephants, but the status quo that in the near future, without artificial means, will see the Asian elephant become extinct. To touch this network as a critical and questioning outsider is to face a wall of denial that is international in dimensions, and to face threats of death, deportation, and imprisonment locally.

The status quo will remain unchanged without full transparency and accountability at all levels of Asian elephant conservation and welfare work. Governmental, corporate and non-governmental funders and financiers of this establishment must be included in a full accounting of why, in spite of billions of dollars spent on conferences, committees, reports, books, films, legislative efforts, and applied research, Asian elephants continue to suffer in captivity and are close to extinction in the wild.

The evident failure of the Asian elephant establishment to conserve elephants in the wild, as distinct from efforts to preserve them in the virtual reality of a zoo or theme park, is part of a much larger failure of the human community world wide to protect biodiversity and human rights, two interdependent elements of sustainability, social justice, eco-justice and world peace. The failure of efforts to keep viable populations of Asian elephants in the wild parallels the failure of the World Bank, recently involved in wildlife conservation efforts, to accomplish its own primary mission to help people. Endemic problems of corruption, lack of oversight, and poor project management and monitoring of initiatives and their consequences are to be acknowledged and rectified. No one likes to have their credibility questioned and to be shown that they have sent good money—even out of US taxpayers' pockets—after bad.

Former World Bank President James D. Wolfensohn blamed these kinds of problems on recipient third world countries like India, but as World Bank economist William Easterly has documented, after 50 years and billions of dollars, the World Bank has made virtually no progress in boosting poor economies and lifting billions of people out of poverty. As Easterly shows, the reasons are far more complex than the endemic problems in recipient countries. There are endemic problems within the World Bank, as with other international governmental and nongovernmental organizations, that are linked with an outmoded Western industrial-economic paradigm; and with those values and perceptions that are the antithesis of those that embrace the sanctity of life, the sovereignty of indigenous peoples, and the sacredness of elephants and all sentient beings.

It is evident that many of the larger nonprofit organizations whose original mission was animal protection, wildlife conservation, and the advancement of human rights and moral progress have gone, like the World Bank, way off course. In the quest for more power and influence through money, they have made a science (called "development") out of mass-mailings, media-campaigns, and multinational corporate associations, forever needing more staff, branch offices and chapters. The executive level salaries that hand-picked Boards of Directors justify and approve from a corporate business perspective of equivalence for a President and CEO of a nonprofit organization, and the glossy publications and annual reports that they applaud, are looked at askance by those working at the grass-roots—their strongest and most loyal constituency and original supporters—whose increasing disenchantment should not be ignored, if not for ethical reasons, then at least for fiduciary reasons. But perhaps, when all is said and done, there are those who believe that there are no real solutions; that elephants will become extinct in the wild, along with indigenous peoples. So why not make a living from "feel-good" humane, conservation, and humanitarian projects even if they believe it is all pointless? And the best we can hope for is to preserve some of the Earth's original flora and fauna in seed and gene banks, and in greenhouses and zoos.

Such fatalism is surely unacceptable and will only worsen the issues that could still be turned around, given the will, the integrity, the vision, effective governance, and the voice of indigenous peoples and all who would give their lives to save the last of what is wild, authentic and true.

Efforts to reduce "human-elephant" conflict by digging trenches around villages and tribal settlements and by erecting solar-panel-driven electrified fences, as advocated by Prof. Raman Sukumar, Chairman of the international WWF/IUCN Asian Elephant Specialist Group, and by the US Fish and Wildlife Service, along with others, are not sufficiently based on in-field knowledge to know the harmful consequences of what they advocate. Solar-cell-dependent electric fences fail when not properly maintained and spare parts are not available. So landowners and tenant farmers growing cash crops connect the fence wires to the main power line (which is illegal, according to Tamil Nadu State Forest Department and Central Government laws and wildlife protection regulations). One tenant farmer was killed in 2002 when he accidentally touched one of his field wires, and local police intervened in an attempt to cover up the cause of death, because the landowner was ultimately responsible for this illegal activity of hot-wiring field fences to kill crop-raiding elephants whose habitat is being encroached and degraded. The US Fish and Wildlife Service and the Performing Animal Welfare Society, in liaison with the International Rotary Club, have put much money and effort out in solar-fence construction in the Nilgiris with the best intentions—but, the harmful consequences of fencing need to be addressed.

Other government and non-government organizations have advocated the Trench solution. In June 2002, a nursing mother elephant in IPAN's area fell into a trench and broke her leg. And there was no equipment available to pull her out—and then what? One tribal elder told us that he saw a large, old female elephant go into a trench so the rest of the herd could walk across her back to reach the fields. The herd was able to help her out of the trench later. The only solution to the human-elephant conflict in the Nilgiris and elsewhere is to remove the human element entirely from farming cash crops that the hungry wild elephants will try to reach.

A proposal to use Kumkhis (trained bull elephants) from nearby elephant camps to assist in keeping wild elephant herds away from crops is not likely to be effective, since there are no vehicles or communication system to bring them at a moment's notice to where they are needed. Also most of these elephants are weak from chronic malnutrition and lack of physical activity since most are kept on short chains 16-17 hours day and night.

Research being funded by the US Fish and Wildlife Service to develop various crop-repellents to deter elephant crop raiding is a paper solution, promising in theory but not realistic in practice, since even simple-to-maintain solar fences are not effectively monitored as to their integrity. Elephants will push trees across electric fences to take them down. Trenches become a trap for other wildlife, e.g., wild boar and deer, and fill up with eroded earth after the rains, and quickly cease to be of any use without responsible community maintenance. It is very unlikely that those communities being impacted by crop-raiding elephants will be sufficiently competent to use the proposed chemical and pheromonal deterrents currently being researched for in-field application. So the assaults of shooting elephants with homemade guns, and blowing up their mouths with explosives placed inside large jack fruits that almost invariably result in slow deaths and great suffering, are likely to continue.

Additional indirect human-elephant conflicts that IPAN has addressed include illegal real estate developments, especially of guest lodges; illegal felling of trees and diversion of streams to irrigate crops and plantations; habitat encroachment and degradation by agriculture and grazing livestock who can spread diseases fatal to elephants and other wildlife.

If certain African countries win their appeal to CITES to kill elephants for their tusks and put the ivory on the world market, they will hasten the demise of Asian elephants in the wild, because ivory poaching will intensify. A *total ban* on all ivory trade is a vital component of Asian elephant conservation.

Funding more field research on elephants that has been going on for the past 30 years in the Nilgiris should be questioned, considering that the precipitous decline in their numbers has not been averted. Indigenous tribal peoples see the wild elephant as part of their culture and future. More should be employed, trained, and appropriately equipped for anti-poaching and forest protection. With better law enforcement, local government accountability, land purchases, relocation of land users from core habitat areas and "corridors," coupled with more open international collaboration and project oversight, there is a chance to save the last of the wild elephant herds. But time is running out, and the end time of Asian elephants in the wild is drawing near.

Since an earlier version of this document has been widely circulated, and following the success of the conservation coalition petition to have the Chief Conservator of Forests in Tamil Nadu not permit the underground Indian Neutrino Research Observatory to be built in the protected wildlife preserve, concerted efforts have been made to re-establish vital elephant corridors and to prohibit further encroachment and construction of guest lodges and settlements. Reducing the competing livestock population is a controversial but essential component of wildlife and habitat protection that remains to be fully addressed.

Conclusions

Some readers will see this report as too confrontational with those involved in elephant conservation and improved welfare in captivity. After all, we had been told repeatedly, we should all work together, in a spirit of cooperation, be reasonable and diplomatic. I (Fox) spoke at Representative Sam Farr's Press Conference in 1999 that was about the US government's decision to lift sanctions against India (that were made when the Indian government violated the nuclear-weapons test ban treaty) by permitting the release of public funds appropriated under the Asian Elephant Conservation Act. At the Press Conference I said that if India, and all who care, cannot help one Asian elephant (Loki) whose videotaped capture and our (IPAN's) documentation of his injuries were shown at this public hearing, then how can we save any elephants? When I shared my misgivings prior to the Press Conference with Rep. Farr (D-17/CA) over the telephone, especially over the need to closely monitor where the money goes and evaluate all funded projects, he said that was not his concern or responsibility. His task, he told me, was to get the appropriated funds under the Act over to India. I surmised that there was pressure from various US corporate business interests, notably Enron, that was at high-financial risk with its investments and commitments in India at that time.

In hindsight, we feel that Loki became a political tool. In India he was politically "hot," according to the Indian Ambassador in Washington, DC, who told Deanna Krantz that the elephant (Loki) was India's second most controversial issue (the first being Pakistan).

IPAN's efforts to help Loki and other elephants at the camp were met with police and Forest Department harassment, *by death and deportation threats*, and by a disinformation campaign that implied that we were making a big fuss over nothing, just for publicity and money, and that the severity of Loki's condition was all a fabrication. This disinformation was spread internationally by the US publication *Animal People*. During this time, Paul Irwin, then president and CEO of the Humane Society of the United States and my employer, claimed that he was told by the Chairman of the Animal Welfare Board of India, and by the Director of the Wildlife Trust, India, that "Dr. Fox would be killed or put in prison if he ever went back to India." I have been back to India several times since this intimation, and have enjoyed nothing but ever increasing local support as Hon. Chief Veterinary Advisor for IPAN. Yet, I was not permitted by Paul Irwin to represent the HSUS, which had initially funded IPAN between 1996-1998, and I was told subsequently that I could "only go to India on my own vacation time and at my own expense." Curiously, Mr. Irwin went with his wife to India around that time, at the behest of the Enron Corporation.

Clearly, the politics of elephant and other animal welfare and conservation issues are complex in the international arena. But the reticence of large organizations with power and influence to be more confrontational and supportive, when animal suffering and species extinction have been thoroughly documented, is surely inexcusable.

Keeping Elephants in Captivity

Our concerns about the treatment and overall well being of captive elephants in zoos and circuses are expressed in the following overview of issues (references on file). Asian elephants, who have such self-awareness shared only by chimpanzees and humans that they can use a mirror as a point of self-reference to groom and preen, are seen by many Westerners as domesticated, intelligent and compliant circus performers and, along with lions and tigers, are the heart of every zoo. So it is no surprise that a national survey in the U.S. in 2006, initiated over the issue of captive elephant welfare in U.S. zoos and circuses, revealed strong public acceptance of keeping elephants in zoos. I contend that their welfare can **never** be adequately provided for, because, even if as adults they are compliant, they are **not** domesticated ,i.e., genetically preconditioned to adapt psychologically or physiologically to the conditions provided by circuses and zoos. If they were, than the many elephant keepers who are killed or maimed every year around the world would not have been harmed by these animals, most of whom are the victims of learned helplessness, and the Stockholm syndrome. Those who do not rebel, and have not been killed, or subjected to the Indian way of 'rogue rehabilitation' (chaining, starvation, and beating, often for weeks, even months in India for those elephants who kill their only too often drunk and oppressed mahouts and Kavali apprentices), die prematurely from broken hearts and spirits that manifest as infertility, depression, impaired immune systems, coupled with the consequences of a life confined to a circus trailer or concrete zoo enclosure—obesity, rotting feet. chronic arthritis, and most probably other diseases similar to those that afflict their captors and exploiters, and those who enjoy seeing elephants do circus tricks, and reach for bananas at the zoo: and who believe that it is educational for children, as well as entertaining. Elephant exploitation and enslavement still has no bounds regardless of nation state, culture, or creed.

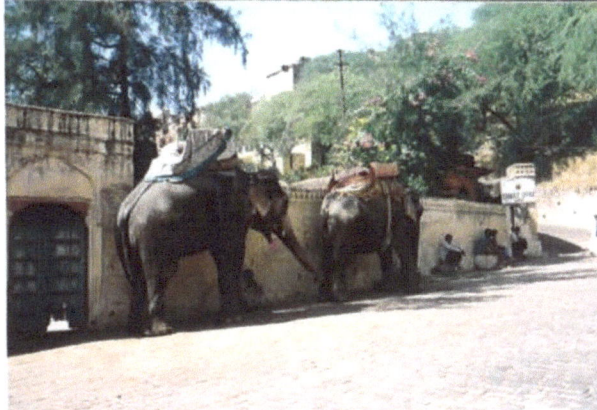

Elephants (above) are exploited for hard labor, such as by the timber industry, often illegally destroying the elephants' habitat; for entertainment as at a State-operated elephant camp, audiences not knowing what training methods have been employed to make them submit and 'perform'. Those below hug the shade in Jaipur, waiting to take tourists on a long ride, day after day.

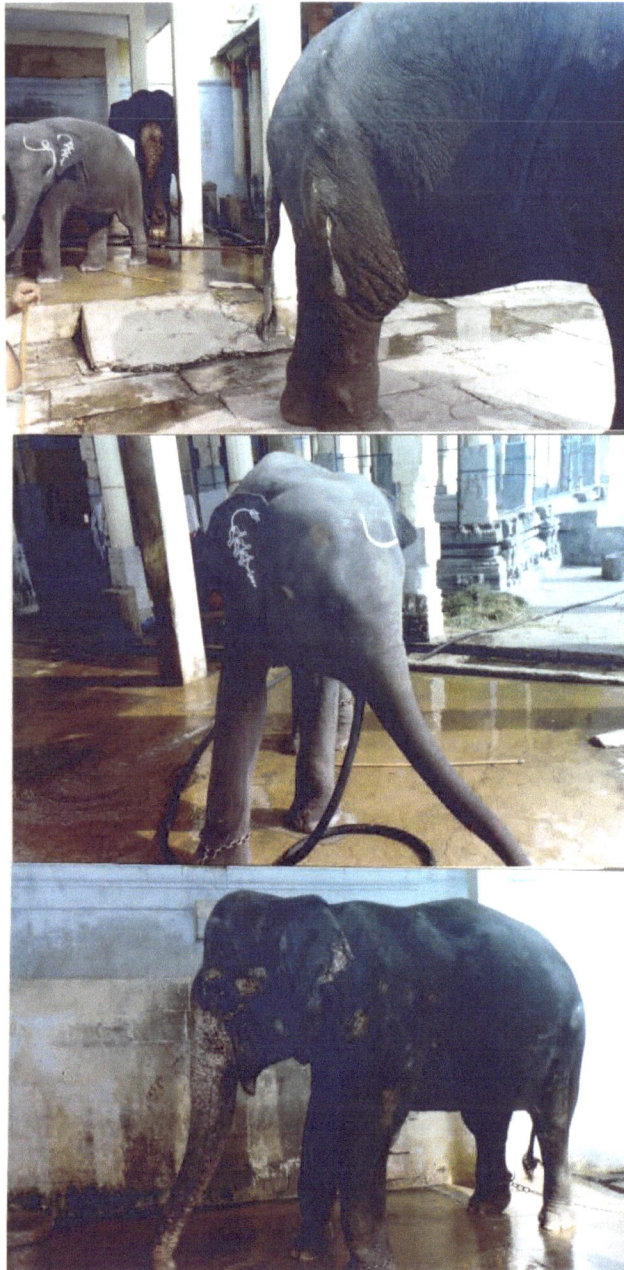

Elephants kept at a temple in Mysore spend their lives in chains and are used for religious ceremonies and rented out for weddings and festivals.

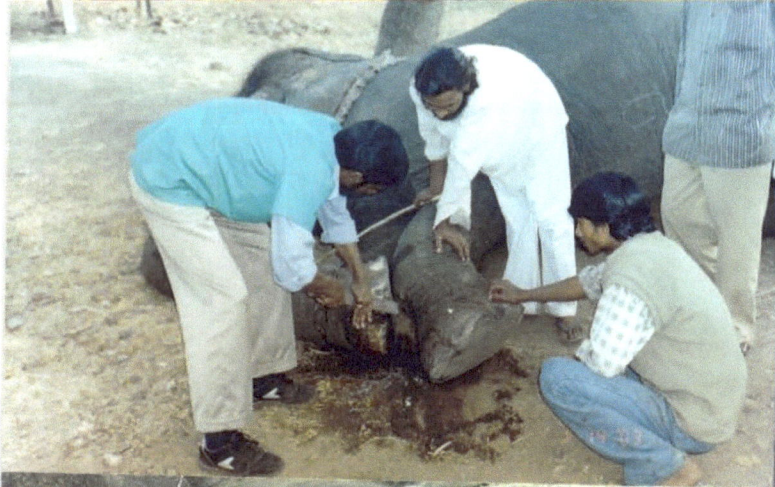

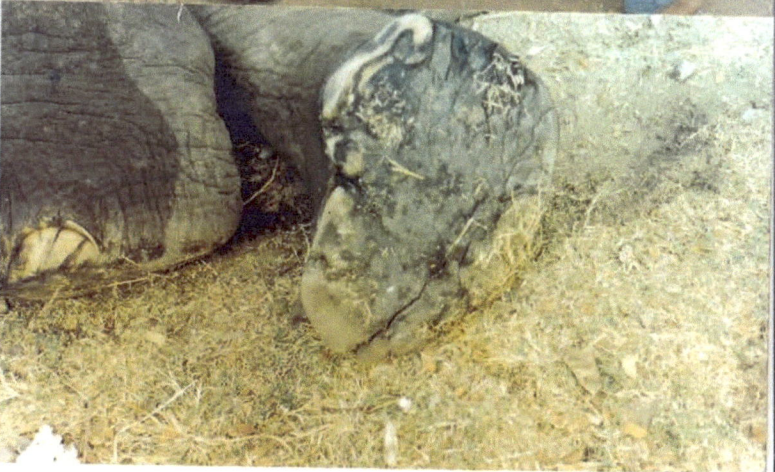

IPAN/USA's team treats the overgrown, infected feet, long neglected, of a temple elephant.

The following overview we prepared as a background document for a law suit filed in 2000 by a coalition of animal protection organizations against Feld Entertainment/ Ringling Bros. Barnum & Bailey Circus for animal cruelty and exploitation of an endangered species. The case was dismissed by the judge on the grounds that the key witness for the prosecution, a former elephant handler employed by Field Enterprises was given money to cover his basic living expenses, during the many months of discovery and preparation of the law suit. Feld Entertainment then sued for damages on charges of bribery, money laundering and obstruction of justice, and took $15.75 million from the coffers of The Humane Society alone, and in December 2012 got $9.5 million from the American Society for the Prevention of Cruelty to Animals.

Captive And Performing Asian Elephants:
Ethical, Health & Welfare Concerns

Along with Asian elephants, the brutal, traditional methods of training them, of breaking their spirits at an early age and forcing them to submit and obey, were exported from India and other Asian countries and adopted by circuses in the West over a century ago. No significant progress has been made since then to improve on these methods of training and handling circus elephants, while some humane standards and procedures have been long in place at better run zoos and wildlife sanctuaries for several years. These standards have been recently revised and upgraded by the AAZPA (Hutchins and Smith,1999, AZA, 2001), some elephant experts conceding that they are neither appropriate nor feasible for performing elephants because of the nature of the traveling circus environment. This fact underlies the growing consensus among elephant experts, including ethologists/animal behavior scientists, conservation biologists, veterinarians, and animal welfare scientists, that elephants do not belong in circuses because humane methods of training, transportation and housing, with provision of an environment that satisfies elephants' physiological and psychological requirements and behavioral and social needs cannot be implemented.

As one Ringling Brothers elephant handler confided to us (and wished to remain anonymous), "Ringling Brothers don't mean to harm elephants—they can't avoid it."

Zoo director David Hancocks (2003), in discussing elephants' needs, and how Asian elephants are cared for in most zoos, concludes that they should not be kept in the average zoo facility but in specially designed sanctuaries. He is adamant that they have no place in the circus and the claimed educational value to children is unsubstantiated fiction. Professor Lori Alward (2003) has similarly reasoned that the circus environment cannot provide for the basic needs of elephants, and that their stereotypic behavior (swaying and rocking to and fro, and other repetitive activities while restrained for many hours, see Gruber et al. 2000), is denied or ignored by the circus industry as a behavioral pathology symptomatic of psychological suffering due to inadequate care and environment.

Several animal studies have shown how environmental deprivation is detrimental to brain development (Diamond et al. 1967), and may actually cause brain damage (basal ganglia anomalies) that is associated with various stereotypic behaviors (Garner and Mason, 20042, Wurbel, 2001). This could be a serious problem with Asian elephants raised with little normal social contact with their own kind, and in an environment that deprives them of what they would experience in the wild, qualitatively and quantitatively, in terms of sensory input and cognitive processing. In order to compensate for such low-level stimulation, animals engage in stereotypic behaviors (repetitive, compulsive actions like swaying, rocking, pacing, rubbing and other behaviors that can lead to self-inflicted injuries), which are self-stimulating, and are a symptom of 'boredom'—extreme environmental restriction and impoverishment—in many species (Wemelsfelder,1990).

Stereotypic behaviors have an obsessive-compulsive, addictive element that has been shown to involve the production of natural opiates to also help relieve the stress and frustration of confinement , and are generally regarded as symptomatic of stress/distress (Marsden, 2002; see also Meyer-Holzapfel 1968, Dantzer 1986, Cabib 1993, Rushen et al. 1993, Lewis et al. 1996, and Garner 1999). To contend that such behavior is adaptive and not a behavioral pathology is to ignore many studies to the contrary, and to deny the scientific evidence that stereotypic behaviors do not evoke some kind of beneficial homeostasis. On the contrary, the bipolar nature of stereotypies (Fox, 1974) that can lead to this erroneous conclusion, which entails their manifestation during reduced levels of stimulation associated with boredom, restraint and frustration on the one hand, and increased levels of stimulation on the other, associated with anticipation of either positive or negative reinforcement (Wemelsfelder, 1990).

As for using negative reinforcement or punishment that circus spokespersons have claimed to be never used, positive reinforcement only being the rule, is documented fiction. The breaking and training of performing elephants follows the traditional "Carrot and Stick" approach (Lenhardt, 2003) entailing the use of negative reinforcement—pain and fear—to make the elephant submissive, coupled with intermittent reward or positive reinforcement (usually a food treat) for good behavior. Zoo elephant handler Maria Gallowy (2003) has shown that female Asian elephants can be successfully trained using positive reinforcement, and that beating and chaining are unnecessary practices in the zoo environment, but that more difficult to handle bull elephants are best managed under protected contact (see also Honeyman et al. 1998, and Laule and Whitaker, 1998). Bulls are not used in circuses.

The high incidence of handlers/trainers being killed or injured by their elephants (22 deaths in 7 years between 1990 and 1997) and elephants becoming "rogues" and going berserk are all a consequence of negative reinforcement, pain, fear, and intimidation. According to Parrott (2000), "You cannot predict when an elephant will explode. Tranquillizers are useless—As long as people are allowed to be in close proximity with elephants, people will continue to be hurt and killed—When an elephant attacks, the difference between a minor injury and death is pure luck." Chaining can result in aggression toward keepers, the practice of chaining increasing the likelihood of keeper injury and death (Roocroft and Zoll, 1994).

In India, according to Dr. Jacob Cheeran (2003), when an elephant is given over to a new mahout (elephant keeper or handler/trainer), the elephant must first be severely beaten so that the new mahout assumes a dominant relationship with the animal, adding that while this is inhumane, it is considered necessary, and that drug-assisted training, i.e. using tranquillizers, should be considered as one alternative. Prof Gary Varner (2003) likens the relationship between the elephant and the trainer who uses pain, fear and intimidation, and intermittent positive reinforcement to the so called Stockholm syndrome where the captive becomes bonded pathologically to the abusive captor.

The use of the bull hook or ankus to cause pain by penetrating the elephant's sensitive skin when used as a spear, or to cause pressure-pain when dug into the skin, coupled with beating with sticks, canes, and shovels, and shocking with electric prods while the elephant is restrained by chains around the legs, is common practice, food and water deprivation often being used to further ensure the animal's compliance, dependence and obedience. Such mistreatment as part of the training of performing elephants sets up a conditioned state of chronic fear and anxiety that is called "learned helplessness" (Seligmen 1975, Overmeier 1981) that shares some characteristics of human depression and despair.

Being raised in an environment of fear and intimidation from an early age, compounded by the trauma of being weaned three or more years too early, having little or no freedom from chains, and being denied the opportunity to freely interact with other elephants, sets up the kind of mind-body disconnect (Fox 2004) that jeopardizes circus elephants' health and welfare. The combination of treatments and conditions results in what is termed psychoneuroimmunosuppression, a brain-body disturbance that has been extensively researched under controlled laboratory conditions (Ader 1981). Some of the clinical signs of this disturbance include increased incidence of infections linked to poor disease resistance, and thus increased morbidity and mortality rates, shorter life-spans, reduced fertility, fecundity and offspring survivability—all common problems in captive and performing elephants and those in poorly managed zoos.

All of these health and welfare problems are predictable since elephants are not biologically or psychologically adapted to and therefore capable of thriving in the traveling circus environment, since there is now irrefutable evidence of bidirectional communication between mind and body, brain and immune system, such that an animal's mood and mental state influences cellular defense and repair mechanisms (McMillan 1999 & 2003, Moberg and Mench 2000, and Panskeep 1998).

An additional factor is the attitude of elephant handlers and mode of relating to their charges that have been shown in several animal studies to influence fertility, fecundity, lactation and overall health (Seabrook 1984, Hemsworth and Coleman 1998). The documented production of "feel good" neurochemicals such as oxytocin, phenylethylamine, dopamine, endorphin, and prolactin in both animals and their care-givers have a profound bonding effect when the attitude is loving and the mode of relating caring and gently playful (Odendall and Meintjes, 2003), and play a significant role in warding off infection and speeding recovery from illness and injury (Fox, 2004).

Stressful treatment, especially rough handling and transportation, of pregnant elephants could have adverse effects on the offspring, while gentle handling both prenatally and postnatally can prove extremely beneficial to the development and health of the offspring, according to several controlled laboratory studies in other mammals (Denenberg 1967, Gellhorn 1968, Fox 1971 and 1986) .

Rearing conditions—the degree of environmental and social complexity—also have a profound effect on brain and behavior development, which are relevant to the well being and psychological health of captive elephants, the majority of whom are raised and kept their entire lives in impoverished environments. Elephants in the wild range 20-25 miles and more in a day, foraging for food for 16-18 hours in their familial home-range, which may cover hundreds of square miles, and also bathing and engaging freely in self-care behaviors such as rubbing, rolling, dust and mud-covering (activities impossible in the circus environment), and have the constant stimulation of their socially complex herd, and of an ever-changing natural environment (Sukumar, 2003 & 1989, Adams, 1981).

Scientists and conservationists recognize that the captive population of Asian elephants in circuses and zoos in the US is not self-sustaining (Weise, 2000). Elephant husbandry/care experts recognize that the captive environment with little normal physical activity being permitted, and having to live on concrete, often contaminated with urine and feces and almost constantly wet, shackled for hours all night and most of the day, forced to travel thousands of miles every year by road and rail, and being physically restrained most of their lives, except when in the ring, to a 2-4 square meter living space, results in obesity, associated infertility and dystokia, and especially in chronic, crippling and painful foot and joint problems (Roocroft and Zoll, 1994). The lack of normal social contact with other elephants results in social incompetence, disrupting the elephants' ability to mate normally, and to successfully raise offspring. "Captive-born elephants are often attacked and rejected by their mothers," states Rees (2000).

Due to the great weight and size of their bodies, elephants that have poor muscle tone because of a lack of regular exercise, are particularly more prone to tendon, muscle and joint injuries, chronic arthritis, bursitis, and rectal and vaginal prolapse especially during training and while performing, for instance, by being made to get up and stand on stools, and when being loaded and off-loaded from trailer cars.

Other health problems due to poor management and the conditions that they must endure include abscesses as a result of trauma and piercing by the ankus, nutritional deficiencies, dermatitis, dental abscesses and caries, colic, enteritis, volvulus, intussusception, internal and external parasites, hepatitis, pneumonia, frostbite, sun burn, and heat stroke. According to Brockett et al. (1999), "indicators of declining health such as foot problems, arthritis and colic, appear to decrease when animals are not chained."

Bacterial infections, many of which are transmissible to humans, include tuberculosis, pasteurellosis or hemorrhagic septicemia, salmonellosis, anthrax, tetanus, E. Coli, and enterotoxemia. Virus infections include foot and mouth disease, herpes virus, papilloma virus, coryza-like syndrome, and two viral diseases that can be transmitted to humans, elephant pox and viral encephalomyelitis. In so far as the USDA inspects performing elephants and evaluates their health in order to protect the public from these zoonotic diseases, former USDA veterinarian and inspector Larson (2000) concludes that "USDA compliance is at best hopelessly ineffective. You should not rely on USDA inspections to provide you with an answer to the problems of circus animal health and care."

The high mortality rates collated by Lambert (1998) in circus elephants—24 deaths between 1994-1998, and at least 31 deaths between 1994-2004 (API, 2004) indicate the seriousness of the performing elephant's plight and their non-viability as a population, since mortalities are greater than the birth-rate in Ringling Brothers circus-elephant replenishing breeding facility (16 offspring produced between 1992 and 2003). Elephants may also sometimes die from stress-related cardiac fibrillation due to fear when having their first training session, or having a new, unfamiliar handler; and die from a "broken heart" following a stressful event such as transportation away from a familiar environment, or loss of a stable-mate (Adams 1981, Schmidt 1986).

This latter observation points to one of many emotional similarities between elephants and humans. The ethogram or behavioral repertoire of both species includes homologous behaviors occurring in the same contexts, that indicate similar emotional awareness and cognitive processing, notably: Maternal and group defense of the young; agitated distress and therefore empathy when a group member is suffering and vocalizing in pain or fear, and attempts to rescue the same; bereavement over loss of a group member, and attempts to touch, lift, carry, and bury the dead. Both species also engage in care-giving and care-soliciting behaviors, including allomothering (baby-sitting), food-sharing, cooperative midwifery, that is, assisting an inexperienced mother to give birth and nurse, and creative social play, including wrestling, chasing, role and dominance reversal, bluffing or deceptive behavior. Furthermore, elephants also employ inventive tool-using behavior involving insight, reasoning and observational learning; self-medication in times of injury and illness; cross- generational transfer of knowledge through observation and example/instruction within the safe nexus of an extended family with complex social relations, communication skills and kinship affiliations. Adult elephants and humans possess "Machiavellian intelligence," being able to understand, predict and manipulate the behavior of others, anticipate the consequences of their actions, and engage in mimicry (Poole, 2003).

While performing, elephant suffering and deprivation are part of the unavoidable conditions of the circus situation, deliberate cruelty in order to dominate, control and discipline these large animals, where males can weigh up to 7257 Kilograms and females 3629 Kg, is considered a necessity for both effective training and safety. Discussing these concerns with officials in India, where elephants are considered sacred, I was repeatedly told that "No one wishes to deliberately harm an elephant, but delinquent behavior must be corrected, and rogues, like criminals, punished" (Krishnamuthy, 1994).

I have seen and documented the wounds on elephants who have been so "so disciplined," even blinded in one eye, and reduced to standing skeletons in shackles after months of being starved for killing or injuring an often abusive handler at India's claimed "best" elephant camp at Theppakadu, in the Nilgiris, Tamil Nadu, South India. But, regardless of the cultural context in which elephants are kept, cruelty is cruelty, and the animals' suffering to the degree documented by impartial observers cannot be condoned or ignored. The means do not justify the ends—public entertainment, profits and jobs for elephant handlers—and can never be condoned by a civilized society with established animal protection, anti-cruelty and endangered species conservation laws. Nor can such mistreatment be rationalized as unavoidable and necessary if elephants are to learn to do tricks and perform in the circus, be safe to handle, be reliable in public places, and to be saved in captivity and not become extinct in the wild.

They are never totally safe to handle in the circus setting; are never wholly reliable in public, because even if captive bred and raised, they are still wild animals; and are not domesticated.

As for helping conserve indigenous populations of Asian elephants in their natural habitats, India Project for Animals and Nature (IPAN) has extensively documented the various projects funded by the US Fish and Wildlife Service, and by Western NGO's, such as digging trenches and putting up electrified fences, and testing pheromone and other chemical deterrents to repel crop-raiding elephants so as to reduce human-elephant conflict and retaliation-slaughter, simply do not work in-field due to a variety of factors. The same holds true for anti-poaching initiatives such as providing vehicles, field equipment and personnel training, but no good guns. The politics of extinction are not being addressed, especially the lack of habitat protection, effective law enforcement (in part due to corruption), and illegal tree cutting, irrigation, and land encroachment.

According to Kreger (2003) of the US Fish and Wildlife Service, the institution in the US seeking to import an endangered species like the Asian elephant must do some "*in situ* enhancement" as part of the agreement. Providing funds for the aforementioned projects that have done nothing in the Nilgiris to achieve any "*in situ* enhancement" is of concern since the exportation of young elephants from a declining wild population in which there are no adult bulls left, is to surely violate the Endangered Species Act by preying on an already at-risk population. Importing "captive bred" young elephants is not acceptable when they have actually been taken from the wild after their mothers have been shot, electrocuted or killed with home-made bombs placed in large jack fruits. The export value of such orphans is a disincentive for the authorities to either effectively patrol, investigate and apprehend elephant killers, or to raise and release the orphans back into the wild. The fatalistic argument that since there is less and less natural habitat left, and too much human population pressure, the only safe place for elephants will be in captivity—where the breeding records are to date non-sustainable, is untenable, when there is still viable habitat and more elephants being born in the wild than in captivity.

Falsified breeding and birthing records aside, what elephants who are born in captivity, as at the Tepakkadu elephant camp (for many years under the direction of the late venerated veterinarian Krishnamoorthy/Krishnamurty, it experienced many problems and deaths, for instance, two young female elephants died from massive tapeworm infestation in August 2004 prior to his own demise), are actually semi-captive bred; the females are allowed into the forest at night on long drag-chains, that cause constant leg problems, in order to hopefully breed with a wild male. The captive bulls at this camp, who are constantly shackled, have no libido, and, along with chained females, are often attacked and severely gored when a wild bull comes through the camp.

 I have met with Western elephant specialists who have been to this camp to collect semen samples, teach the staff the use of ultrasonic pregnancy diagnostics, and to set up a micro-chip identification scheme of individual elephants, activities that stretch the imagination when claimed to have any relevance to elephant conservation in the wild. Preservation, *in vitro*, as cryo-processed as freeze-stored sperm, eggs, and cells, and, *in vivo*, in circuses and zoos, is closer to the truth.

Another truth that is denied by those who exploit elephants for commercial purposes is that elephants are being studied, researched and "managed" into extinction. Those protective of their vested interests in maintaining captive elephant populations for commercial purposes, regardless of the PR-hype of funding reproduction research and successful attempts at interventive and invasive reproduction-enhancement, as with artificial insemination, are actually accelerating the demise of the species. They are encouraging the taking of infant elephants from the wild, and are shifting the focus from concerted and effective conservation *in situ*, to preservation in captivity, which cannot be justified scientifically and ethically, or sustained biologically. When all elephants are gone from the wild, those in captivity will soon join them, for in captivity they neither belong nor can sustainably reproduce. The same can be said for the great mammals of the sea such as the Orca and dolphins who also manifest self-awareness and empathy, qualities which seem to be lacking, collectively, in the human species.

What Future For The Largest Remaining Wild Population?

India's largest population of Asian elephants in three southern Indian States of Kerala, Karnataka and Tamil Nadu is in grave danger, if initiatives based on a recently approved "science-directed" conservation proposal, under the chairmanship of ecologist Raman Sukumar, PhD from the Indian Institute of Science, Bangalore, are not examined and serious flaws corrected. Some scientists are linked with government and nongovernment organizations and vested interests and monopolize how funds are allocated and priorities determined when it comes to the conservation of endangered species, and the protection and restoration of their habitats. This monopoly has a mafia-like ethos when scientific authority is aligned with special interests detrimental to animals' well-being and species' survival. This is evident in the *'Report of the Karnataka Elephant Task Force'* submitted on September 2012 to the Honourable High Court of Karnataka, and which has now been approved and is being put in to action. (http://www.academia.edu/2628781/Report_of_the_Karnataka_Elephant_Task_Force.pdf) Following our critical analysis of these conservation proposals (see below) contained in this Report to save the Asian elephant, we call for greater transparency, accountability and responsibility by authorities, including foreign donor agencies and enlightened philanthropists who understand the connections between human and animal health and well-being and ecosystem preservation and restoration.

Critique of the September 2012 Report under the Chairmanship of Professor Raman Sukumar

Many of the proposed initiatives to secure the future of the possibly over-estimated population of 6,299 elephants in these three southern states (an estimated 12% of a total Indian population of 28,000, of which 4% are captive in Karnataka) may prove effective if adequately funded and monitored. But the report lacks essential time-line data on population numbers of elephants versus human and livestock numbers over the past few decades—humans and their encroachment, fuel wood, water and livestock needs being the major issues if elephant habitat is to be given life-saving CPR—conservation, preservation and restoration.

This Task Force Report does not address the evident conflicts and killings of elephants by elephants, indicative of a population under great resource stress associated with habitat loss, focusing primarily on ways to reduce human-elephant conflicts and killings, electrocution being the most prevalent cause of death. The report's proposal to capture elephants in high conflict areas and keep them in captivity for Forest Department use met four dissenting voices on the expert panel whose statement of opposition (Appendix 1) was castigated with vehemence and ridicule by Sukumar, perhaps based on his own arrogance, fears and fallacies. It should be noted that in 2009 Sukumar, ironically a recipient of several lucrative and prestigious international conservation awards, actually supported a major mining project for the proposed construction of an underground Neutrino Observatory research laboratory in the heart of the Nilgiri Biosphere Reserve, and adjacent to the Mudumalai Tiger Reserve, an important elephant habitat. This would have been extremely harmful to wildlife and was successfully blocked by the well documented environmental impact concerns of local and international NGOs, which Sukumar sought to discredit, also (for details go to http://www.nilgiribiospherereserve.com/articles/154-environmentalism/733- open-letter-to-dr-raman-sukumar-ino-project.html).

The report advocates capturing and removing elephants where there is high conflict with humans. Having secured video documentation of injurious, high-risk, near-death stressful wild elephant capture in the Nilgiris and witnessing the breaking of their spirits so they can be of Forest Department use, the euthanasia of the entire herd as a last-resort intervention in high conflict areas may be argued to be a compassionate alternative to their capture and Forest Department "domestication." But, killing is ethically unacceptable when there is little evidence of effective habitat restoration. One recent review of the consequences of translocating "problem" elephants concluded that such translocation actually defeats the goals of both human-conflict mitigation and elephant conservation (go to http://www.plosone.org/article/info%3Adoi%2F10.1371%2Fjournal.pone.0050917). Where capture and relocation to suitable habitat is not feasible, and with the likelihood of surviving members of the herd returning to their original location, captive sanctuaries with minimal human contact need to be established, rather than simply replenishing Forest Department elephant 'camps' (where captive breeding is minimal and elephant suffering rampant) with wild-caught animals. The use of Kumkis, trained bull elephants used to separate mothers from their calves in the camps, for handling elephants that go berserk from a life in chains, and for the capture of wild elephants, should become something of the past. With new chemical tranquillizer delivery systems, appropriate restraints, lifting and transportation equipment, the art and science of elephant capture can be improved to minimize injuries, capture myopathy, suffering, and death.

Kumkis can cause serious injuries in the process of capturing wild elephants.

The Karnataka Elephant Task Force in its report identified two regions—Alur and Tumkur district's Savandurga—as "elephant removal zones" for the "unacceptable levels" of conflict. The report said, "Here, a herd of 25-30 elephants inhabit a tiny 5-sq km forest patch circumscribed by agricultural fields that they routinely raid." According to State Forest Department statistics, between 1986 and 2011, elephants killed 46 people and injured over 240. The Department employs a combination of methods that incorporate elements from traditional khedda (stockade traps) and chemical tranquilizing to remove a target number of 25 elephants from these zones.

By early May 2014, 22 elephants had been removed and put in to State operated elephant camps according to news.webindia123.com. (For video documentation of capture using trained bull elephants see 'Capturing a wild elephant in Savandurga.' *Suvama News*, YouTube on Videos of karnataka capture of wild elephants bing.com/videos). It should be noted that in such human-elephant conflict zones there is human retaliation. Elephants are often killed by high voltage wires set at their head-height around fields; die slowly when shot in the flank with home-made guns; succumb to pesticide poisoning in sweetened food baits, and from pressure-bombs placed in large jack fruit that explode in their mouths.

It is noteworthy that Irula and Kurumba tribals in Tamil Nadu informed us that 50 or more years ago they could walk without any danger and lived at peace with the elephants, knowing the names of close to 100 bull elephants in the now designated Nilgiris Global Biosphere Reserve where adult bulls are rare.

Provision of fodder and constructed water troughs in designated elephant and other wildlife habitat during the dry season would do much to reduce human-elephant conflicts when the animals leave their preserves in search of food and water. Dozens of elephants have died in Karnataka in 2012 due to lack of water during the dry season (http://www.wrrcbangalore.org/index.php/elephant-deaths-in-karnataka-2012-india).

Effective containment of elephants in designated zones and improved barriers and deterrents to protect crops and settlements will lead to more 'elephant islands' and increased inter-elephant conflicts and deaths from starvation, fighting over finite resources and reduced fertility and fecundity. The flowering and death of bamboo across India beginning in 2010, a major food source for elephants, is an additional stress factor and trigger for crop-raiding by starving herds.

The assertion in this report that establishing one particular elephant corridor, which many elephant conservationists are calling for, is not needed, because the two currently separated elephant groups are two discrete, independent entities, is *patently absurd* considering the historical range of the species. No mention is made of the need to open the Tirunelli-Kurakote elephant corridor in Kerala, which the World Land Trust is seeking funds to restore.

Mother elephant who took 2 months to die from gunshot wounds. The herd came and took her 2- year old calf away with them.

Most disturbing is the statement that designated Elephant Conservation Zones "could also include a certain number of human settlements." Without better protection of these Conservation Zones, this Report will amount to a blueprint for the gradual extinction of the Asian elephant in the wild in southern India, if the continued deforestation, livestock grazing, irrigation projects, quarrying, and spread of guest lodges for tourism are not stopped, and the settlers not relocated—and if, in so called Co-Existence Zones, sustainable agricultural practices, zero-grazing and ecoforestry practices are not initiated. Leases on Forest Department lands by rubber, cardamom, tea, and coffee estates should not be renewed when they expire, and new guest lodge construction prohibited.

Young bull elephant killed by a jack fruit containing a pressure bomb that exploded in his mouth.

Illegal harvesting of bamboo in elephant corridor takes away their food source.

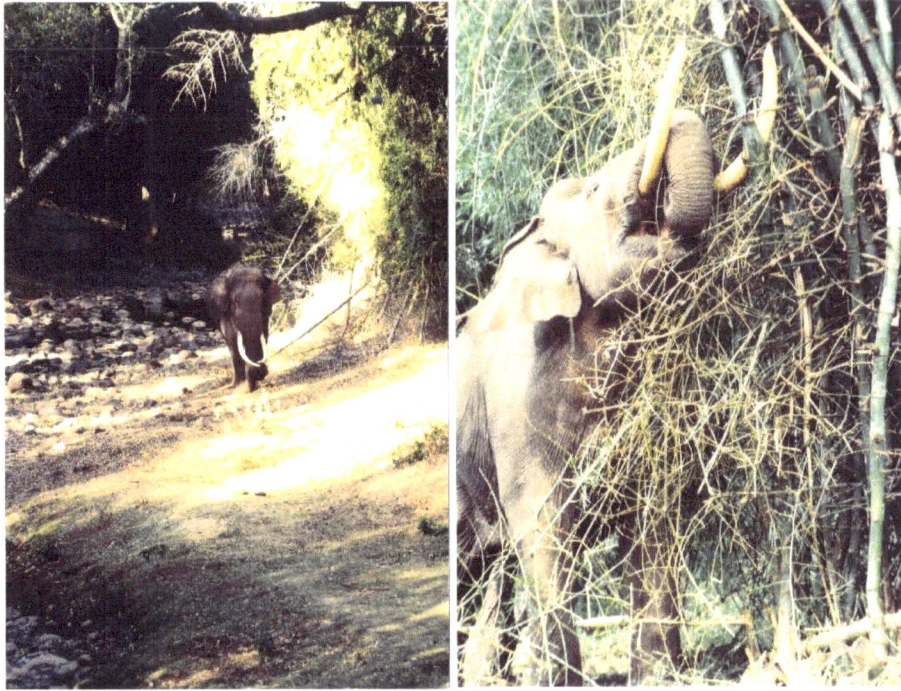

Tuskers in the Nilgiris 30 years ago. Now no adult bulls are left in the Mudumalai Wildlife Preserve.

Photo documentation of elephants, electrocuted, shot, and poached in the Nilgiris between 1989-2004.

The in-field Senior Veterinarian for India Project for Animals & Nature/USA, over a 9-month period in 2008-2009, documented 13 elephant mortalities in the Nilgiris Global Biosphere Reserve in Tamil Nadu. These figures alone are grounds for grave concern and immediate remedial action. This veterinary colleague provided the following synopsis of the critical status of elephants in this bioregion:

One. The herds are fragmented and many small herds are facing problems without the matriarch in addition to the poaching even of immature males for tusks.

Two. The calf mortality is on the rise because the herd (mothers) has to move faster than in earlier time in search of fodder and water, the young calves often losing their way and meeting with tragedy. We can witness very small groups to the level of two elephants in many places. These are the crop raiders and live close to the human habitat. These small groups will hide in the nearby jungle during the day time and enter the cultivated lands in the night just for their food.

Three. Wild elephants have been breaking into carrot and cabbage crop fields for food for the last our to five years which we never heard of before 2010.

Four. The elephant habitat is shrinking due to increase in human activity in the name of plantations and for the settlement of Sri Lankan Tamil repatriation by a political party to win votes and to establish its presence in the very elephant and wild life habitat. Thousands of hectares of grassland as well as shola forests were destroyed during the 1980's for this purpose.

Five. Exotic noxious weeds such as *Lantana camara*, *Eupatorium* and *Parthenium* are becoming abundant and are major menace and contaminant of wild fodder. There have been attempts at removing these weeds but more needs to be done since it affect forage for elephants and other wildlife.

Elephant calves separated from their mothers (often killed) cannot survive.

Six. Tourism is also one of the causes for the elephant-human conflicts. During 1980 there were two private guest houses in Masinagudi, which is one of the crucial elephant and wild life habitat areas, and now there are around 120 private guest houses that attract thousands of local and foreign tourists every month. These privately owned guest houses are located on the elephant corridors. Such intensive tourism pressure, along with expanded and intensified agricultural activities has created a severe water shortage and with less and less for wildlife. Many rich people purchased land and built their bungalows in the corridor area, Raman Sukumar being one among them. The so called conservationists or government authorities are not able to implement a plan even today to compensate the agricultural community living in the core area of Mudumalai wild life area and relocate them as was proposed during 1985-86.

Captive calf, mother electrocuted, died after a few days in State elephant camp enclosure.

Too many orphan calves starve to death from inadequate care.

Seven. In the case of wild life and human conflicts the victims are principally the under-privileged and the wild animals; no rich person is affected so far. The local poor farmers are doing their usual, traditional cultivation but the scarcity of food in the jungle due to mismanagement is the major cause of elephants raiding their crops.

Eight. Elephants even break open houses and eat the families' provisions. This was unheard of 5 years back. It is due to habitat fragmentation and loss of matriarchal leadership.

Nine. To conserve the present population both the conservationists and the government should be very transparent and pay fair compensation to the land owners based on the extent of their land holding instead of "golden hand-shakes" to the large land holders. Some people who had a little land such as 0.25 to 0.50 acres received rupees 10 lakhs each and give up their land. But those holding 1 to 5 acres and over 10 acres are refusing to accept the compensation and instead demand the same acreage of land in another location which the above-said two parties are not able to implement.

Ten. In my opinion it is very late to implement an acceptable land-recovery for wildlife and further delay will lead to the extinction of the elephant. A very acceptable and reasonable compensation will be the only option along with ensuring the availability of habitat and fodder for the wild elephants instead of more conferences and publications.

Authorities were unable to remove this elephant, who died mired in deep mud created by a combination of "Check-dam" construction for crop irrigation and the illegal removal of rocks from the stream-bed for building purposes.

Dr. Ullas Karanth is another Indian wildlife scientist like Sukumar who is a recipient of international funds and awards for conservation, specifically for tigers, but he recently stated that water should not be provided for elephants in the dry season as it would interfere with their natural population dynamics (http://www.thehindu.com/news/national/tamil-nadu/water-scarcity-may-increase-mananimal- conflict/article5884482). What is so natural about our protected areas is a question that Karanth must answer. When tourist resorts and agricultural fields have access to a continuous supply of water, why not the elephants? About 54 elephants died in the dry season of 2013 due to lack of water in Kabini reservoir in Karnataka, and it was said to be 'normal' (http://www.thehindu.com/news/national/karnataka/it-will-soon-be-paradise-regained-for-wild-elephants/article5747166.ece; http://www.wrrcbangalore.org/index.php/elephant-deaths-in-karnataka-2012-india/).

India has done a remarkable job protecting her wildlife and their habitats against the relentless pressures of rising population and insatiable appetites for natural resources by the rich and poor alike. She has been a shining light to the world for millennia of the ethics and spirituality of co-existence, non- violence and respect for the integrity of the natural world which is a national heritage in sacred trust. The complexities of conservation efforts on behalf of whatever endangered species and communities— be they elephants, tigers or tribals—are challenging enough without being clouded by the narrow assertions of scientific opinion or derailed by political influence, corruption and conflicts of interest. I have no doubt that most members of this *Karnataka Elephant Task Force* are genuinely concerned about the fate of elephants wild and captive. How such concern is translated beyond words into appropriate and effective action will soon be a matter of history. As the Scottish poet Robert Burns quipped, "The best laid plans of mice and men oft go awry."

The Asian elephant in India faces an uncertain future surviving within marginal habitats in densely populated landscapes with increasing anthropogenic impact and ivory poaching (http://www.sciencedirect.com/science/article/pii/S1616504713000463). The Indian population has undergone a bottleneck due to extensive ivory poaching in the 1990's and 2000's which has resulted in highly skewed sex ratios (http://www.asesg.org/PDFfiles/Gajah/32-06-Baskaran.pdf). The elephant population in the Nilgiris-Eastern Ghats range, although the largest in the world, was found to have low genetic diversity (http://www.nature.com/hdy/journal/v94/n1/full/6800568a.html). Therefore, the loss of individuals will cause further genetic erosion of the population.

India's Sate and Central governments' efforts in designating and protecting wildlife habitat and biodiversity against the pressures of population expansion, poverty and increasing demands for land and natural resources need to be redoubled. The ancient cultural traditions of Asian elephant exploitation throughout their entire range also needs to be examined from the perspectives of enslavement and initiating economically viable alternatives for those whose livelihoods elephants have supported and enriched for generations, and for others whose crops and lives are threatened by hungry and thirsty herds. The vested interests of the American and European circus industry in replenishing its declining captive Asian elephant population with replacement calves taken from the wild and also from India's elephant camps, along with sperm from captive bulls for artificial insemination under a bogus umbrella of funding elephant conservation, need to be questioned as well as the continued exploitation of elephants by this entertainment industry.

Like the "Shame of wearing fur" of a wild animal is a call to conscience to consumers in the West, the "Shame of riding on an elephant" and "Shame of the elephant circus" (an outmoded cultural tradition at best when the animals' spirits have been broken and herd-life and freedom taken away) are clarion calls from all four corners of the world. The capturing of elephants to protect people in high conflict areas will only escalate if the juggernaut of human numbers and habitat encroachment is not arrested. Capture and captive breeding under the guise of species conservation when there is no future possibility of reintroduction into the wild should not be the fate of the Asian elephant.

Compassionate respect for all life, giving elephants and all nonhuman species equal and fair consideration as to their fate, is a first step toward effective resolution of human-animal conflicts and establishing the kind of bioethical democracy that is the foundation of a viable and truly civilized society. Then all will speak in the one voice that comes when the rights and interests of non-human animals are embraced equally as our own.

Postscript

Sacred Elephants

In the racial memory of elephants
Long before the dawn of human consciousness,
Perhaps they, like us much later,
Recognized and revered a Buddha
Or a Krishna in their midst:

Appearing in their own image
Once in every millennium
To guide, inform, inspire,
Awakening them to the nature of divinity,
The divinity in their nature,
And the power of loving kindness.

If this is true for us,
Then their truth we share
To celebrate the sacred presence
That illumes our mortal lives.

Perhaps elephants are more enlightened,
Buddha-natured, Krishna conscious
Than we, a younger species
Beginning to evolve, recognize and revere
To save them from extinction
And all that makes us human,

Giving loving kindness to them all
And every creature great and small.

Michael W. Fox, July 2011
Inspired by Heathcote William's book, *Sacred Elephant*.

"Bumpy" one of the last of the Nilgiri Tuskers, killed by poachers in the early 1980s.

Authors' Note: On 5 March 2015 Feld Entertainment, owner of Ringling Brothers and Barnum & Bailey Circus, who had been using elephants in their touring circuses in the U.S. for no less than 145 years, announced that over a 3-year period they would phase out using elephants in their three traveling circuses that go each year to 115 cities. Company representatives attributed their decision, according to Vice President Alana Feld, to "a mood shift among our customers." Also, tours were difficult to arrange with many cities and counties having passed anti-circus elephant ordinances. Kenneth Feld, Chairman and CEO, said the company made this "unprecedented" decision to focus its elephant work on conservation programs to save the species from extinction. The circus will send 13 elephants to the Center for Elephant Conservation in central Florida.

CHAPTER 5

Sustainable Farming & Conservation: Indian & Global Perspectives

IPAN   **Agriculture And Wildlife: Creating Harmony And Economic Security**

This report is in four parts, providing an overview of sustainable agricultural practices (Part1). These provide core solutions for the problems detailed in one of the world's richest biodiversity "hot spots" in the Nilgiris, South India (Part 2). Diseases spread from domestic animals to wildlife are addressed (Part 3). Photo documentation of these interrelated concerns affecting wild and domesticated animals, biodiversity, the local economy and well-being of indigenous peoples is provided (Part 4).

Part 1: *Sustainable Farming Practices*

The goal and standard of "sustainable" agriculture is ecological soundness, meaning that farming enterprises do not deplete local natural resources---water and top soil quality and quantity. Everything is recycled in one way or another—crop residues serving as feed for a few farmed animals whose manure is composted and used as fertilizer, or for biogas, and even urine as a crop spray to control pests and as a fertilizer. Their manure is not shipped out for sale as a fertilizer, as done in the Nilgiris, S. India and elsewhere, which depletes local soil quality. Synthetic chemical fertilizers, herbicides and pesticides are used minimally, and when not used at all the sustainable farming enterprise qualifies for Organic status.

No matter how small, in terms of amount of land and animals, each family-owned sustainable farming enterprise, with a few goats, sheep, poultry and milk cows, is integrated with a community of other smallholdings which together make for one large cooperative to optimize the production of a diversity of products from milk to feed for animals. Each small-holding feeds its own family, contributes to feeding the community through the local market and reaches more distant markets through the co-op to generate the income needed for feed, seed, animal replacements, and occasional hired labor.

Generally a few ruminant animals are kept for milk—dairy cows, buffalo, goats or sheep—sometimes being grazed free-range, the different species grazing in rotation helping to maximize forage utilization and manuring of the land. Chickens or other domestic birds may free-range with them to help control ticks and other harmful, disease-transmitting insects. Humane treatment and husbandry practices are paramount, including herders (guard dogs in some areas), and night corrals to deter predators and facilitate manure collection for recycling. Sustainable farmers do not engage in wildlife habitat encroachment, illegal irrigation, deforestation, or predator control activities such as setting out poison bait or using snares/snoozes. Their farm animal husbandry practices minimize predation and the need for predator control.

In conversations with old farmers, for instance, one by the name of Linga Gowder, he told us he never sought to retaliate when a tiger took on of his calves—that was his rental payment for grazing in the jungle. Another we have on film, showing his gouged thigh scar caused by a panther whom he drove off his cattle in the night corral, but did not blame, saying he was on the panther's land. Another retired farmer told us that in the old days there were never 'weeds'—they had no word for weed—and they never needed to use chemical fertilizers and pesticides.

Alternative 'zero grazing,' where fresh-cut forage, hay and supplements are brought to the animals, calls for them to be humanely confined, with no over-crowding, proper housing with adequate shade/shelter, fresh water and feed-trough space to avoid bullying, and well drained ground surface that is kept clean with dry lying-up areas. Zero-grazing has become the ecologically unsound modus operandi of CAFOs— Concentrated/Confined Animal Feeding Operations—also called factory farms. But when applied to small numbers of dairy cattle and other farmed animals, *zero grazing helps prevent the spread of diseases* between domestic animals and wildlife in particular, which is inevitable in free-range/extensive livestock operations.

The presence of a veterinarian is of the utmost importance for not only the health of animals but also for instructing and educating about humane treatment of animals. All of this was virtually non-existent before Deanna Krantz and I started our free veterinary services in the Nilgiris under the banner of India Project for Animals & Nature (IPAN). The importance of maintaining a healthy population of on-farm and free-roaming village dogs through rabies and other vaccinations, treatments for various parasites and spay/neuter to reduce overpopulation helps prevent the spread of several diseases from dogs to livestock, such as rabies, echinococcosis, leptospirosis and neosporosis. Addressing the health and welfare of dogs (and cats where prevalent) is therefore an essential contribution to sustainable farming practices where these different species come into close contact.

Male offspring of milk-producing animals are yet to be treated humanely, be they fed and slaughtered locally, or subjected to castration (often by pounding their testicles between rocks), branding-identification, and transportation to other farmers who raise them for deployment as draft animals or for their meat and ultimate slaughter. The horrific forced march to slaughter and being beaten and crammed into over-crowded trucks and railroad cars resulting in great suffering, injury and death is India's shame and must become something of the past. Spent dairy cows from private and government dairy factories should be slaughtered locally rather than being put in "gowshallas," holding facilities until they die from disease, starvation and general lack of proper care for economic reasons, the income from their manure, hides and bones not withstanding (see Chapter 3, *Holy Cow: The Sacred & the Suffering*).

Advocates of intensive, so called high-input farming systems see the sustainable farming movement as a step back in time that will not feed the hungry world. But in fact it is a step back to recapture millennia of traditional wisdom and integrate such knowledge, as well as seed and breed stock varieties, with environmentally less damaging food production practices. Two cases in point in the Nilgiris: More emphasis on rain-fed crops and consumption of same, such as highly nutritious ragi (a kind of millet) rather that rice, which needs constant irrigation and is processed into low-nutrient white rice; reduction of 'scrub' cattle numbers kept as manure producers and grazing in wildlife habitat, competing for feed and spreading disease and their replacement with more productive hybrid, locally adapted dairy cattle. Goat herds allowed to free-range with a protective herder, kept for milk and meat contribute to the local economy and also to forest and land-management, consuming brush and controlling invasive plants as an alternative to herbicide applications.

Tragically, millions of rural families in India and other developing countries, including some who are forced to move into urban slums, are in part sustained by an often shared milk cow and by a few goats and chickens. Few are fortunate to have access to grazing land, much communal land being overgrazed as per ecologist Garret Hardin's predicted global "tragedy of the commons." Without urgently needed subsidized veterinary services and supplies of animal feed and water (rather than scavenging off garbage) this sector's farmed animal population (including bovids and equids used for draft work) continues to be a neglected yet significant incubator for diseases harmful to other farmed animals, to wildlife and to the human populace locally, nationally and internationally.

Sustainable agriculture means going forward with an economically and ecologically more viable, cost-effective, healthful, socially just and even regenerative approach to feeding ourselves and domesticated animals, while minimizing the adverse impacts on wildlife, many species of which play an indirect role in facilitating ecologically sound farming practices as by controlling crop and livestock pests and diseases.

The humane and sustainable production of farmed animal produce has its roots in traditional, small family farming enterprises and is the antithesis of industrial agricultures factory farms. Claimed as the way to feed expanding urban populations in developing countries and to meet the rising demand for meat with rising incomes, these factory farms cause widespread pollution (along with tanneries), especially of drinking water; contribute to climate change; cause food born and zoonotic diseases (notably swine and avian influenza), and contribute to world hunger by indirectly impoverishing and marginalizing the rural poor, displacing their once sustainable, local and regional cooperative farming enterprises. According to the UN Children's Fund 2013 report, 48%—61.7 million—of India's children below the age of 5 are physically stunted and are mentally and immunologically impaired. Yet ironically India has become the world's leading exporter of beef (from buffalo) now leading Argentina and the U.S.

There is a finite amount of land and water to feed and process increasing numbers of factory farmed animals, a hidden cost compounded by agricultural petrochemicals and animal pharmaceutical products from antibiotics to growth stimulating hormones, all having harmful public health and environmental consequences. In the best of worlds this calls for a phasing out of factory farms and more enlightened dietary choices based on the principles of humane sustainable agriculture as still practiced to varying degrees by traditional indigenous farmers, as in the Nilgiris, Tamil Nadu, South India. But they, along with the wildlife in those bioregions where various and diverse cultures have flourished for generations, are endangered by the threat of spreading factory farms and non-sustainable agricultural practices, including the adoption and proliferation of GMOs---genetically engineered varieties of commodity crops, that take the food from the mouths and land away from the ploughs of local people.

For India, a re-awakening of the spiritual tradition of vegetarianism (which means veganism for the more affluent who can afford high protein non-animal foods to help reduce the suffering and often slow starvation of spent dairy cows in gowshallas), and not becoming one of the world's leading exporters of meat and hides, is enlightened national self-interest; and more widespread respect and support of humane and sustainable farming enterprises, along with land reform, would do much to turn back the tide of rural poverty and crippling consequences of infant malnutrition.

In sum, the ethics of sustainable agriculture call for respect for the living soil, for water and air quality and for the health, welfare and quality of life both workers and all animals in the farming enterprise.

The ethical demands and hallmarks of agricultural sustainability are in the spirit and cooperative economy of mutually enhancing relationships between the land and those who farm it or engage in other commercial activities while accommodating the interests and rights of indigenous peoples, wildlife, especially endangered animal and plant species, and the natural environment.

PART 2: PROBLEMS IN & AROUND THE NILGIRI BIOSPHERE RESERVE

Neither elephants, tigers and other wildlife nor indigenous peoples will have a secure future in the UNESCO designated Nilgiri Biosphere Reserve (details below) until sustainable agricultural and forestry practices are promoted and adopted in this bioregion, and animal and environmental protection laws are enforced.

Thanks to an information network of villagers and tribal peoples who had intimate knowledge of the jungle, and who came to trust our in-field CPR (conservation, preservation and restoration) work in the Nilgiris that began in 1995, we were able to document and report to the appropriate authorities various illegal activities in the Global Biosphere Reserve, which the local people could not do for fear of reprisal, even death.

These illegal activities included: construction of guest lodges, tea shops and temples, and operation of a brick factory in restricted areas; land encroachment for agriculture; grazing livestock in restricted areas; illegal diversion of rivers and streams into private lands and pumping water from same to irrigate cash-crops, and pollution of same by agrichemicals and small industries; expansion of eucalyptus, tea and coffee plantations that seriously depleted and variously polluted the water table; opening up of new roads and illegal improvement of forest roads in the Reserve restricted for use by Forest and Wildlife Departments only; illegal cutting of trees for firewood and lumber; non-sustainable, destructive harvesting of forest products (mosses, lichens, gooseberries, soap nut, tamarind, etc.); movement of un-quarantined, un-inspected and infected livestock into the Reserve; sport hunting, poaching, and sale of wild meat, skins, elephant ivory and other wildlife products; killing elephants and other wildlife with homemade bombs and by shooting, electrocuting, poisoning, and snaring ('snoosing'); illegal removal of cattle manure and the quarrying and removal of rocks and sand; burning of the remains of killed elephants and endangered wildlife guar, tiger, leopard by forest staff to avoid punishment for not catching the killers; procurement of tribal girls for prostitution at local guest lodges, and the killing and suicides of same; misappropriation of foreign funds, which were provided to empower tribal women and to facilitate family planning and economic security, but were used to bribe officials, purchase land and vehicles and to build a guest lodge for eco-tourism; illegal receipt and misappropriation of foreign donations and government funds to operate a bogus animal shelter and refuge that provided no free services to the local community as mandated by its Charter of Incorporation; bribing of various government employees, including high-ranking government officials, roadside check-post officers, police and SPCA animal welfare inspectors, veterinarians falsifying livestock numbers, vaccination records and autopsy reports on wild and domestic animals; providers and purchasers of contaminated and inferior grade food for captive elephants falsifying receipts, and elephant caretakers falsifying body weight, injury and treatment records.

The Indian Institute of Science (IIS) enabled a German student to bring several hives of domesticated bees into the Nilgiri Reserve to study their behavior and adaptability, an extremely irresponsible project that could endanger the wild bees through contagious infections and competition and endanger the traditional sustainable economy of the Honey Kurumbas. Lead IIS elephant researcher Raman Sukumar, winner of international conservation awards, supported a proposed mining project in the Biosphere Reserve (located in the heart of the Tiger conservation area) to create an underground research facility for neutrino physics research, such activity causing potential harm to this ecosystem and wildlife. IPAN and other concerned local organizations convinced the Forest Dept. authorities to reject this proposal. IIS scientists have been doing research studies on elephants and other wildlife for decades in the Nilgiris, accidentally killing some, e.g., tranquilizing elephants for radio- collaring. They have shown no evidence of any active and effective involvement in species and habitat protection and restoration at the local level where it is clearly needed and involves cooperative efforts with local peoples and government officials. Doing research for research sake, and not getting involved in the politics of CPR, as elephants and other endangered species are on the brink of extinction, is like Nero fiddling while Rome burns.

Other government and non-government projects like tree planting, introduction of 'improved' water buffalo, and growing feed and fodder for local livestock have failed over the years due to poor management, lack of oversight, misappropriation of funds, and limited if any consultation with village leaders and tribal elders whose wisdom is rarely appreciated.

Other major problems included increasing human incursion and settlement with increased pressure on water and fuel wood resources; the accidental introduction of a highly invasive weeds in imported crop seeds; the abandonment of growing traditional, rain-fed and highly nutritious varieties of staple crops like ragi for local consumption; a high population of 'scrub' cattle raised primarily as manure producers (for out-of-state sale as organic fertilizer) and which are extensively grazed, become extremely malnourished, and suffer terribly during the dry season, and which compete with wildlife and spread disease.

PART 3: DISEASES SPREAD BY DOMESTIC ANIMALS TO WILDLIFE;
RISK-REDUCTION IN THE NILGIRIS BIOSPHERE RESERVE.

Diagnosis, treatment and prevention of various communicable diseases in domestic animals are integral and critical aspects of wildlife protection and conservation, especially in those areas where there are small communities of people and their animals living in close proximity to viable populations of wildlife, as in the Nilgiris, Tamil Nadu, South India, and where it is not feasible to translocate these people and their domesticated animals, or to effectively police the grazing of livestock in protected wildlife habitat.

Since 1996, India Project for Animals and Nature has been addressing this issue in the U.N. designated Nilgiri Global Biosphere Reserve, where the population of "scrub" or "nondescript" cattle (kept mainly for manure collection to sell for fertilizer), water buffalo, sheep and goats, greatly outnumber the human population. Low-income farmers cannot afford the necessary veterinary services for sick animals, especially for low-value scrub cattle, or to purchase dips and sprays to control ticks that can spread various diseases. Because many of these animals are not only heavily parasitized, but are also chronically malnourished, particularly during the recent prolonged droughts, their immune systems are compromised. This can mean that the pathogens that they may carry become more virulent, according to controlled laboratory studies that have demonstrated a relationship between an animal's nutrition and bacterial pathogenicity.

As a consequence, IPAN's free veterinary services are in great demand, and our in-field work has provided a consistent monitor of disease problems in livestock that are of public health and economic concern to the community, and put wildlife at risk. Also since IPAN is often called by the Govt. Forest Dept. to perform autopsies on elephants, guar, sambar, cheetah, wild dog, wild boar, leopard/panther, and tiger, to attend to sick and injured wildlife, and to capture and vaccinate dogs against rabies during epidemics, we have generated much in-field information about wildlife diseases. They are most often contracted from domestic animals from the tribal and village communities in and around the State-operated 450 sq.km Mudumalai Wildlife Sanctuary (which is one of the major protected preserves in the Nilgiri Biosphere Reserve). This conclusion is based on the fact that it is *after* disease outbreaks have been seen in the domestic animal population that they are seen in the wild, rather than before.

IPAN also assists in Govt. vaccination programs, inoculating livestock against four major diseases that can be transmitted to wildlife, and have been diagnosed in wildlife by IPAN, namely, *Foot and Mouth Disease, Black Quarter, Anthrax, and Haemorrhagic Septicemia.* These diseases have been seen in elephants, and guar or Indian bison. Only an estimated 40 percent of the indigenous livestock are actually vaccinated, in part because of inadequate infrastructure. Lack of transportation, effective refrigeration/cold-chain to preserve the vaccines, and inaccurate records of numbers of animals being kept, and receiving vaccinations, mean that these and other endangered species are at risk in the Nilgiris. *Canine Distemper* is a common disease that IPAN controls by vaccinating village dogs, this highly communicable disease (since dogs scavenge and hunt in the wildlife preserve) putting the tiger, panther, wild dog, jackal, wild cat, hyena, mongoose, Palm civet, and otter at risk. These and other wildlife species are also affected by *Rabies*, so IPAN also vaccinates dogs to control this disease that not infrequently spread to people and livestock by rabid dogs. *Parvovirus disease* has also been diagnosed and treated in village dogs, and is regarded as a potential zoonotic infection for susceptible wild carnivore species. IPAN's spay/neuter program has significantly reduced the dog population in several communities, and this initiative, coupled with educating dog owners about proper nutrition and with regular worming and vaccinations, has greatly improved the welfare and wellbeing of these animals, and significantly reduced their risk to wildlife.

But the risk to wildlife posed by the high density of cattle and other domestic livestock (including poultry) in the bioregion continues, with the following zoonotic diseases being recognized: Tick-born *Babesiosis, Ehrlichiosis,* and *Theileriosis,* Contagious *pustular dermatitis, Brucellosis* (Contagious Abortion), *Pasteurellosis* (Hemorrhagic Septicemia), *avian Newcastle disease, Paratuberculosis* (Johne's disease), *Tuberculosis, Sarcoptes* (mange), Dermatophysosis (ringworm), *Facsioliasis* (liver fluke disease), *Tetanus, Anthrax, Leptospirosis, Infectious Keratoconjunctivitis* (pinkeye), *Foot and Mouth disease, and Black Quarter.*

*One of many village laying-hens, dying from Newcastle disease,
also highly contagious to wild birds.*

An epidemic of *Rinderpest* in livestock spread to the wild ungulates, killing thousands in 1985, eliminating 90 percent of the Guar in the Nilgiris, with a serious knock-down domino effect on the tiger, panther and wild dog populations. There are undoubtedly other zoonotic diseases that have yet to be identified. Prolonged periods of drought, that mean starvation for herbivores because of a lack of vegetative growth (much of which is taken by livestock or cut and harvested to feed to same), coupled with limited drinking water sources due to hydroelectric power projects and diversion of streams for irrigation, mean a weakening of animals' immune systems, and greater susceptibility to disease, especially when they forced to share grazing, drinking and wallowing areas with unhealthy domestic animals.

Many of these diseases, like Foot and Mouth disease, Black Quarter, and Mange lead to *Myiasis,* where skin lesions are invaded by the flesh-eating maggots of parasitic flies, resulting in great suffering in afflicted animals. These parasitic flies and ticks proliferate on sick and injured scrub cattle, and are a significant threat to wildlife of all species that might sustain even a small cut on their bodies.

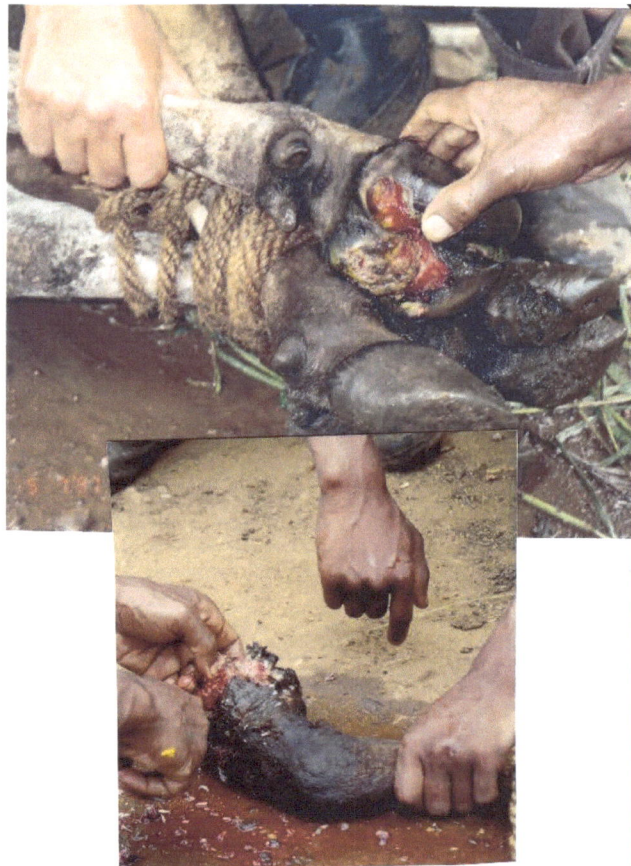

Cattle and other animals afflicted with Foot & Mouth disease cannot eat because of ulcerating lesions in their mouths, or walk and get to water with lesions in their feet; note maggots being removed from rotting cow's foot.

IPAN has also documented pesticide poisoning (with organophosphate and carbamate-type insecticides) of wildlife, put out in retaliation against predators (panther and tiger) and crop-raiders (elephant, wild boar, deer), on whom urea is also used as a poison. Vultures and other scavengers of poisoned livestock carcasses are also victim to such retaliation. IPAN addresses this aspect of wildlife protection by reporting all such fatalities (and also those caused by electrocution, snares/nooses, shot-guns, and home-made bombs placed inside Jack fruit) to the authorities: and provides veterinary certification of predator-loss by doing autopsies on farmers' animals, so they may be compensated by the Forest Department; then they will have less incentive to seek retaliation. Unfortunately the compensation process is erratic at best.

IPAN's veterinary efforts to prevent, treat and eliminate these zoonotic diseases is coupled with teaching livestock keepers basic animal husbandry, care and nutrition. IPAN attempts to reduce the scrub cattle population by castrating all scrub bulls and by providing a more valuable hybrid bull whose offspring would produce more milk, a more profitable product than cow manure from scrub cattle.

In summary, IPAN has shown that working with the people and focusing on zoonotic diseases in domestic animals is a necessary and effective approach to wildlife protection and conservation that could well be adopted in similar bioregions where the "conflicts" between people and wildlife remain to be resolved.

It is extremely short-sighted, and evidence of flawed conservation strategy and policy, for the health and welfare of domestic animals in and around wildlife habitat not to be given top priority, along with poaching and land encroachment and degradation. To not fully address zoonotic diseases spread to wildlife by domestic animals, and not take effective steps to reduce the overall domestic animal population, as IPAN is doing in the Mudumalai Wildlife Sanctuary and National Park section of the Nilgiri Biosphere Reserve, is to guarantee the failure of Project Tiger, Project Elephant and other conservation efforts. In the adjoining state of Karnataka's Nagarahole and Bandipur National Parks three elephants died of anthrax in March 2004, an outbreak associated with an estimated 200,000 cattle that graze in and around these wildlife preserves. State Govt. Veterinary Services are ill-equipped, under-funded and under-staffed to meet these challenges. IPAN has helped rectify these in the Nilgiris for over eight years, working in close cooperation with the State Forest Department, farmers and herders, and community leaders, with former State veterinary service officer Dr. M. Sugumaran, now IPAN's full-time veterinarian, and our dedicated local staff.

Part 4: Photo-Documentation Of Ecological & Animal Health & Protection Issues From the IPAN Archives

Adult bull elephants are now a rare sight in the Nilgiris due to habitat encroachment, being killed for their ivory and for raiding crops.

Illegal pumping of river water for crop irrigation in wildlife protected area.

Illegally established small plantations and grazing of domestic buffalo in protected wildlife habitat reflect deficiencies in law enforcement in the Nilgiri Biosphere reserve.

Occasional massive deforestation for commercial timber and land clearing for construction of guest houses have caused great harm to this Biosphere Reserve.

Extensive grazing of "scrub" cattle in and around the Biosphere Reserve for decades has competed with and harmed wildlife and plant biodiversity and should be prohibited.

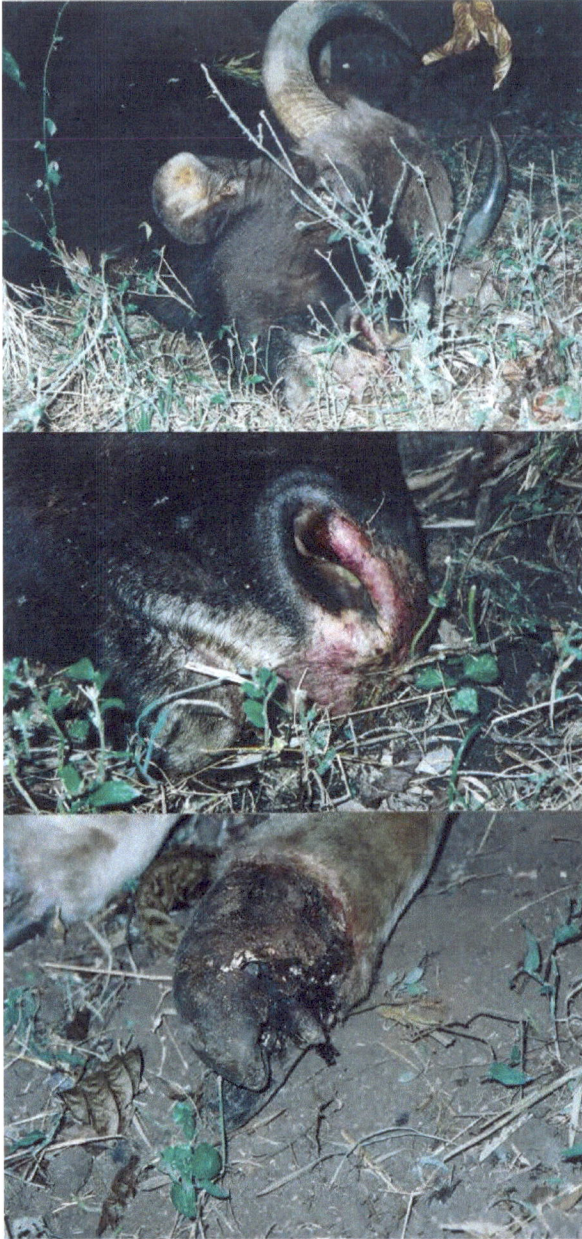

Free-grazing cattle spread diseases to wildlife, including foot-and mouth disease, which killed this Guar or Indian bison, an endangered species. Note lesions on mouth and hooves, resulting in death from being unable to walk and feed.

Retaliation by cattle herders against predators has decimated the tiger and leopard populations: Tiger killed with poison bait and leopard caught and suffocated in a wire snare.

While removal of collected cow manure for export and sale as fertilizer is prohibited in the bioregion, it is a daily early morning activity. Spent, non-productive cattle endure long marches to slaughter, legal only in Kerala, many collapsing en-route.

IPAN provides essential veterinary services to help prevent livestock diseases, alleviate suffering and improve animal care.

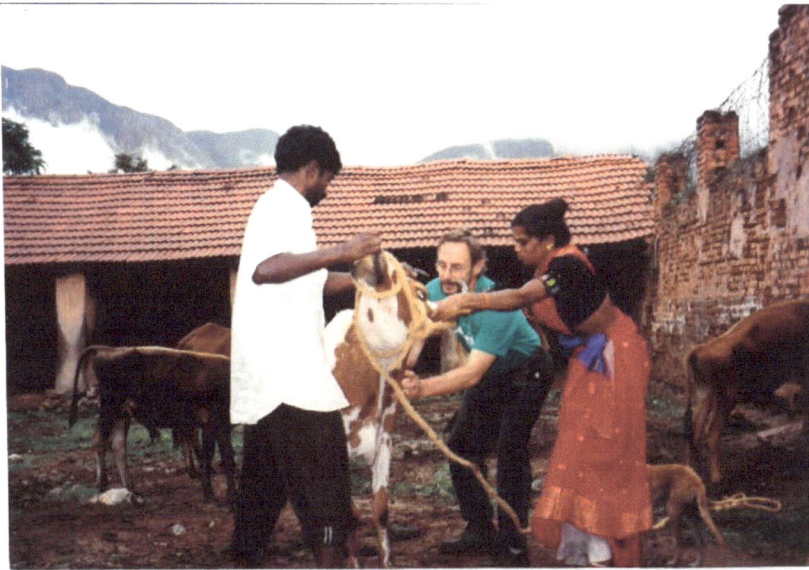

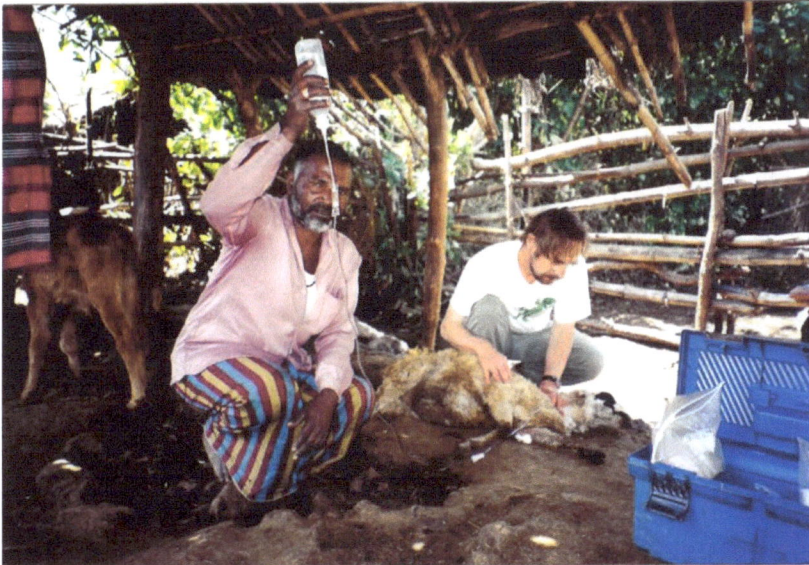

Veterinary obstetrics secures safe birth of a genetically improved hybrid calf so fewer cows are needed for milk production.

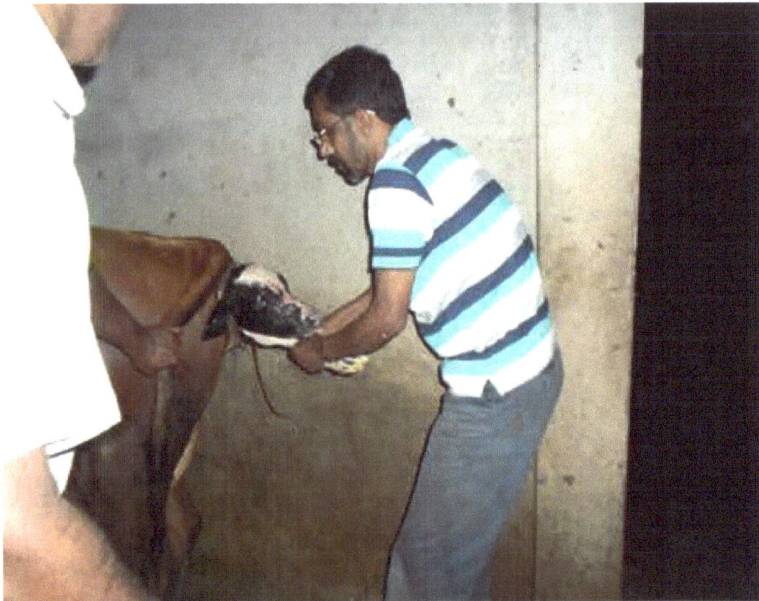

Jersey hybrid cow crossed with a local breed nurses her hybrid calf.

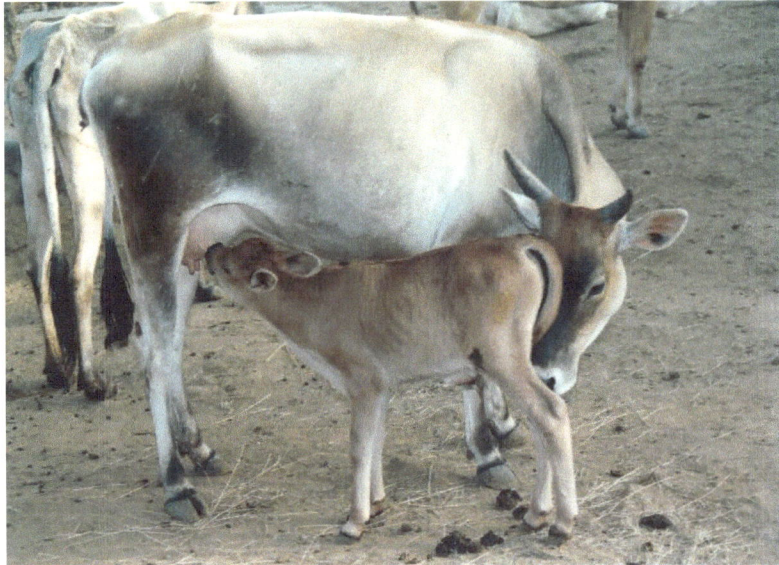

Neutering and vaccinating village dogs helps control their numbers and prevent spread of disease, especially rabies which they spread to livestock and wildlife.

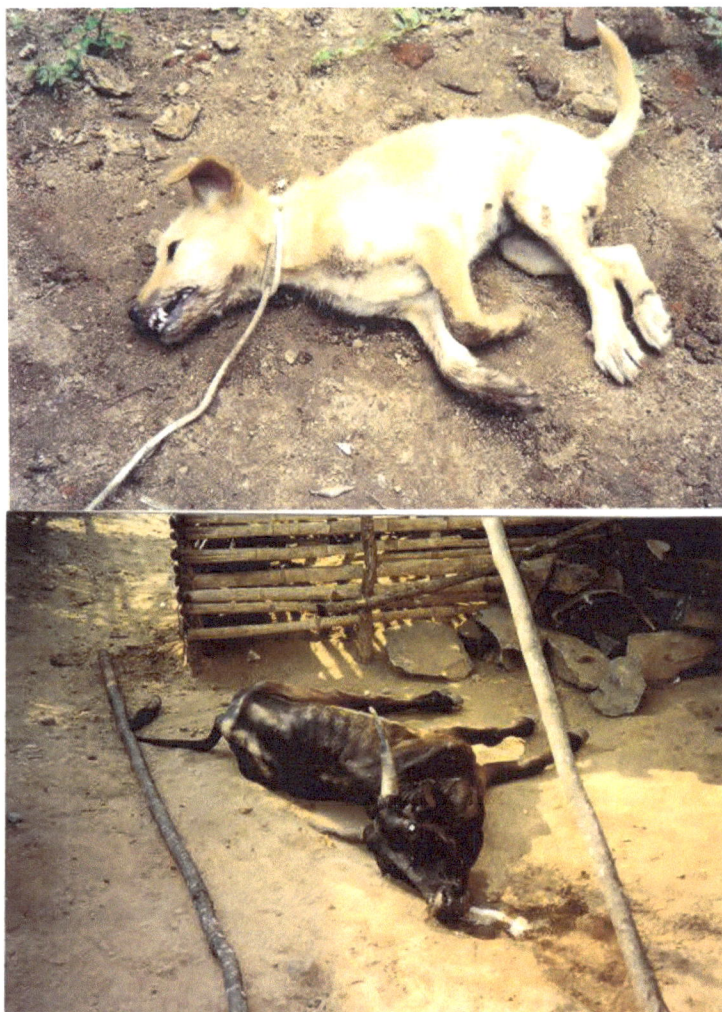

Rabies can spread rapidly from community to community in the absence of effective veterinary preventive vaccinations. Young dog and old bullock in terminal seizure-stage of this disease.

India's expanding coal industry in response to the nation's energy deficit—some 300 million people live without power—and to spur the economy, with growth predicted to soon outpace China's, is going in the *opposite direction* of international agreements to reduce the threat of greenhouse gas emissions on the world's climate. Brokering assistance and technology from America to build nuclear reactors as a purportedly safer source of energy may founder if U.S. companies do not have sufficient protection from potential liability in the case of mishaps to justify their investment. Deforestation for fuelwood, regional water shortages and water wars, and lack of sanitation and food security for a burgeoning population, cannot be ignored or given a lower priority to industrial growth.

Widespread use of insecticides and herbicides, especially Roundup, around the biosphere reserve, puts all life at risk, notably in streams entering the biosphere reserve, possibly responsible for this fish kill.

Progress at the expense of wildlife, biodiversity and the rights and interests of indigenous peoples is ethically unacceptable, as per this IPAN community presentation by Dr. Fox concerning imported products such as Roundup herbicide and GMOs---genetically modified seeds and crops.

*Founder & Director of IPAN, Deanna Krantz, tending to Ramu,
a straying, abandoned bullock, in recovery from surgery by
her team to remove an infected eye smashed with a nail-club
for taking vegetables from a village stall in Masinagudi.*

CHAPTER 6
Animal Values & The Dimensions Of Suffering

Voices for Animals

Many good people have written eloquent, heart-felt words to express concern for animals and to inspire their protection from human ignorance, cruelty and indifference, especially over the last three centuries. During this time, however, animal suffering has increased and intensified everywhere in the world, especially as the result of industrial-scale exploitation and the loss and annihilation of species and habitats. Regardless of moving appeals for compassionate action and respect for all life, there has been a veritable holocaust of animal use and abuse. This means that the 'voices for the voiceless' continues to fall on deaf ears, to be either unheard or even ridiculed by those with vested interest in protecting—not animals—but the profitable status quo of their exploitation.

The pioneering biological scientist Charles Darwin wrote: "Love for all living creatures is the most noble attribute of man" and as a reminder he would write on his hand "Not superior." Before him, Leonardo da Vinci, who abjured the consumption of meat, opined "the time will come when people such as I will look upon the murder of animals as they now look upon the murder of men." While the late Pope John Paul 11 asserted in an address before a gathering of veterinarians, "It is certain that animals were created for man's use," he subsequently declared in a public audience in 1990 that "also the animals possess a soul and men must love and feel solidarity with our smaller brethren," and that all animals are "fruit of the creative action of the Holy Spirit and merit respect." In fact, they are "as near to God as men are."

The late Cesar Chavez, President of United Farm Workers of America, with whom I (Fox) shared the podium at an animal rights conference in California, posted this statement on December 26, 1990 to Action for Animals in Oakland CA. He is a rare social reformer to link violence toward humans and other animals with the long-overdue revolution of non-violence toward all sentient beings. He wrote: "Kindness and compassion toward all living things is a mark of a civilized society. Conversely, cruelty, whether it is directed against human beings or against animals, is not the exclusive province of any one culture or community of people…. Racism, economic deprival, dog fighting and cock fighting, bull fighting and rodeo are cut from the same fabric: Violence…. Only when we have become nonviolent towards all life will we have learned to live well ourselves."

Today there is no unanimity between different cultures and nation states as to how we should treat other animals who, like us, are sentient beings, and what duties we have to preserve or facilitate their well-being. While in most societies there are individuals who care deeply for animals, their well-being is undermined by economic priorities in all nations rich and poor. Profit and investor-driven animal industries, notably large scale factory livestock farming and fishing, and in the third world, wildlife poaching (for example, antelope for bush meat, elephants for their ivory, rhinos for their horns, and tigers for their bones), and inadequate veterinary services for family-sustaining livestock, broken beasts of burden and ever-multiplying community dogs, mean an exponential increase in animal suffering over the past few decades. Human population expansion and increasing affluence, with the rising demands for meat and other products of animal origin, combine to create an almost-invisible epidemic in wildlife extinctions, habitat destruction, and climate change. Improvements in the care and welfare of domestic animals and captive wildlife, and efforts to protect threatened species are overshadowed by the intensified expansion of animal exploitation.

The rights and interests of indigenous peoples who strive to live in traditional and ecologically sustainable ways need greater respect and legal standing. Protection and restoration of cultural and biological diversity go hand in hand. But as all cultures must either evolve or perish, including indigenous people whose proclaimed traditional native rights to harpoon whales, slaughter dolphins and to kill wolves and other threatened species for ceremonial purposes have questionable bioethical validity in this modern age with its twin ecologically devastating crises of overpopulation and over-consumption.

The continued international trade in increasingly rare species for the pet trade and animal collections, zoo and folk medicine markets, and the trade in furs from captive-raised and live-trapped animals, call for prohibition rather than tighter regulations and enforcement. Wildlife having to 'pay its own way' by generating revenues from safari trophy hunting, like the lions of Tanzania, and the wolves of North America, from the sale of recreational 'sport' hunting and commercial trapping licenses, should be questioned when wildlife management practices are closer to farming than to enhancing ecosystem health and optimal biodiversity. So called 'canned hunts' of indigenous and exotic, imported wild species raised on private ranches, and from surplus zoo and collector stock, are anathema to humane sensibilities.

The commercial entertainment value of wild animals exploited by circuses and zoos, and erroneously claimed to be of educational value, is surely less than the animals' intrinsic worth and right to live normal lives in their natural environments. The commercial and scientific value of animals, including the transgenic (genetically engineered) and cloned, used in research as human disease models continues to overshadow progress in public health, disease prevention and the adoption of non-animal research alternatives.

Just as the ecological value of indigenous wild species, and of humanely and sustainably managed farmed animals is being more widely recognized, so the judicial system is beginning to accept the emotional value of companion animals. A greater cross-cultural appreciation of the value of animals as healers and teachers would do much to elevate animals' legal and moral standing. However, the subjective, traditional cultural values of animals—totemic, symbolic, esthetic, social, and spiritual—are being increasingly discounted and ignored politically and ethically, and marginalized by the objective, materialistic, quasi-scientific and economic values of the dominant industrial-consumer society. But, with greater recognition and adoption of the principles of One Health, which links human health with animal and environmental health, the value of healthy animal populations, wild and domesticated, may yet become a significant public health priority. The value of animals as indicators and managers of healthy ecosystems and as controllers of zoonotic diseases—such as bats reducing insect populations and snakes keeping rodent numbers down—cannot be denied.

A redoubling of civil society, with humane, animal protection and welfare initiatives, nationally and internationally, is urgently needed, and should be high on the agenda of the United Nations.

These initiatives must include more effective and informative public outreach, mass-media engagement, the commitment of public officials, legislators, educators, universities, religious leaders and institutions, and the private, corporate sector. Human health and economic security are *dependent* upon environmental quality, optimal ecosystem biodiversity and animal health and well-being. Realizing these 'One Earth-One Health' connections to inspire concerted international action to improve animal health and well-being, and conservation, preservation and restoration of wild lands—as per the mission of organizations such as the World Vets (www.worldvets.org), Veterinarians Without Borders (www.vetwithoutborders.net), International Animal Law (www.internationalanimallaw.com) and the Institute for Global Health & Health Policy (www.ighhp.org)—is *enlightened self-interest*. It is as important as public health, human population and consumption control, food security, and air, soil and water quality.

Which Are The Most Animal-Friendly Countries?

The animal welfare charity World Animal Protection (WAP, formerly known as the World International Society for the Protection of Animals), which has offices in 14 countries, posted an interactive *"Animal Protection Index"* on their website (www.worldanimalprotection.org), ranking some 50 countries on its own animal protection standards. These are based on the following criteria: Formal recognition of animal sentience; Support for their Universal Declaration on Animal Welfare; Laws against causing animal suffering; Protecting animals used in farming, animals in captivity, companion animals, animals used in scientific research; and the welfare of wild animals.

A top grade of 'A' was given to the United Kingdom, Switzerland, Austria, and New Zealand. However, I was stunned when they gave India a higher rating ('C,' which was also given to Sweden, France and the Philippines) than Canada and the U.S., which along with Japan, Pakistan, Italy, and Indonesia, were given a lower rating of 'D.' Having visited and lectured in most of these countries on animal welfare issues, and also having spent several years supporting efforts to improve the plight of animals in India, I see yet another major international animal protection organization wasting its time and donor moneys on yet another questionable mission. On the surface it seems laudable, even promising to help reduce zoonoses (animal-to-human diseases), but WAP papers-over the terrible plight of animals in India which I have been documenting for decades. Even the most enlightened laws and declarations are meaningless when corruption, disinformation and lack of enforcement are not addressed.

The real value of any living being, be it a tree, a whale or a wolf, can be better appreciated with a scientific understanding of their ecological fitness. Trees are more than firewood and lumber; whales more than oil and meat; and, wolves more than trophies and fur coats. Our ecological, economic, social, emotional, cultural, and spiritual dependence on domestic and wild animals has a history more ancient than any existing civilization on Earth today. When we can separate our thinking from all such dependencies and associated values, and have a more objective, impartial understanding of the biology and ecological purpose of all living beings, we will better appreciate their inherent value. Only then we may generate the necessary bioethical principles to help govern and guide all our relationships with the living, sentient community of planet Earth for the greater good. Animals help us think and make us human.

Understanding the intrinsic value of animals in particular, and their instrumental value as contributors and indicators of ecological health and biodiversity, can help us place reasonable ethical limits and legal constraints on potentially harmful human values, purposes and demands that are placed upon the animal kingdom. Human progress may then be measured by a necessary decrease in the suffering of animals, and with compassion as the compass of civilization. The word humane will become synonymous with being human. For Charles Darwin, as reflected in his book *The Descent of Man*, this would indeed be an evolutionary leap for *Homo sapiens*. From the perspective of Albert Schweitzer MD, "Until he extends his circle of compassion to include all living things, man himself will not himself find peace."

Much suffering we bring upon ourselves, our families, communities, and other animals, when we are not mindful of the consequences of our actions. It is a challenge to live as harmlessly as possible in a culture of consumerism, where life is treated as a commodity along with Nature's physical resources. While natural disasters, many aggravated by various human activities, along with famine and pestilence, take their toll, we humans and other species who share this Earth with us will continue to be victims in this whole quantum field of human-caused suffering from generation to generation until there is veneration for all living beings, plant and animal, who express and sustain the life and beauty of our planet home. Their well-being is integral to our own in body, mind and spirit.

Tibetan Buddhist monk the Ven. K. C. Ayang Rinpoche blessing deer in Nara Park, Japan, July 1985, and a 'miracle' rainbow of Buddha-light of compassion was caught on camera.

The Earth will be more secure when all children are educated and inspired to regard and treat all creatures great and small as original blessings, as feeling beings—even rats show empathy to other rats—who have a place in the wheel of life, some as companions, healers, teachers, and many others as co-creators, helping maintain a healthy environment for us all. Such tender regard is the foundation for bioethical sensibility, which is the guiding light for a sane and civil society and more viable civilization.

There is a quickening of chaos and suffering in the world and of humankind's awakening. This collision of dark and light creates the spark of human self-realization that can catalyze our evolution as a species, and our revolution as a global community to become pan-empathic in relation to the sentient living community of planet Earth. We then become reconnected with all that births and grows, feels and dies, loves and knows—every leaf, tree, forest, whale, wolf, and sunburst-singing lark.

Note
Deanna L. Krantz founded in 1995 and directs the India Project for Animals & Nature (IPAN/USA), a project of Global Communications for Conservation, New York, a non-profit organization. Michael W. Fox serves as Chief Veterinary Consultant for IPAN/USA, which provides much needed free veterinary services coupled with wildlife and habitat monitoring in this bioregion (Krantz and Fox receive *no remuneration* for their services). Email IPAN@erols.com for more information.

ADDENDUM
What Happened To IPAN?
(India Project For Animals & Nature)

Based in part on Michael W. Fox's earlier field research* in the last of the wild jungle in Tamil Nadu, South India, recently designated the Nilgiris Global Biosphere Reserve by the United Nations—a precious region of unique cultural and biological diversity—Deanna Krantz proposed and established IPAN there. IPAN was created in 1996 as a project with Global Communications for Conservation, New York. IPAN's vision was to improve the health and welfare of domestic animals in this region so that the wildlife would also benefit from fewer diseases being spread by sick dogs, cows and other livestock, and in the process help improve the health and economy of the indigenous people in the many villages and tribal communities in this bioregion.

IPAN was subsequently registered with the Indian government as a charity in 2002, with IPAN's Founder, Deanna Krantz as co-Managing Trustee along with her husband Dr. Michael W. Fox. Nigel Otter, a local small dairy farmer, who was serving as Deputy Director and Manager of IPAN's Animal Refuge at his 3-acre farm, was named Trustee/Chairman.

In establishing a presence in the Biosphere Reserve by operating an animal shelter and refuge and by offering free veterinary services to the surrounding villages and more remote tribal communities, Krantz was positioned to more effectively monitor various foreign-funded projects focused on helping the poor tribal and village communities, and to protect wildlife, including endangered species such as the Asian elephant and tiger; and to document and report illegal activities, from land encroachment and tree-cutting, to stream diversion for crop irrigation, and the slaughter of elephants, tigers and other protected wildlife. She also enforced anti-cruelty and wildlife protection laws, exposing the cruel transportation of livestock by the 'cattle mafia' that was in collusion with the police, and the 'land mafia' that was in collusion with the State Forest Department. In the process she became a voice for the indigenous Tribal peoples who saw their traditional way of life being obliterated, along with the last of the wild.

Krantz first became involved with animal welfare and conservation issues in the Nilgiris in 1996 when she was invited by the Nilgiris Animal Welfare Society (NAWS) to restore their defunct animal shelter and Donkey Rest Home. There she lived and worked for no salary for some two years. She raised funds, trained staff, hired a veterinarian and won the support and respect of the local community. But the Managing Trustees of the NAWS, who were all business men and members of the Jain religious sect, under pressure from local vested interests and whose inhumane and illegal activities she had embarrassingly uncovered, eventually got her evicted under court order. They had never expected Deanna to stay and work, and they wrongly assumed that she would, like most visiting foreigners, simply send them money for the animals so they could continue to run their facility that was essentially a death-camp for spent cows (where they were essentially starved to death and given no veterinary care) and a week-end country retreat for affluent members of the Jain community.

It was then that she became involved with a local scrub-dairy farmer, Nigel Otter, when he offered her the use of his 3-acre Hill View farm, where he kept some dairy cows and calves, as a refuge for some one hundred animals including a herd of some 90 donkeys, 9 which were being bred and sold into hard-labor) that she took with her when she was evicted from the Nilgiris Animal Welfare Society's 120 acre animal refuge, across the river from Mr. Otter's small -holding.

Krantz went on to raise sufficient funds from various donors to provide for 16 staff, plus medical and surgical supplies, field equipment, feed for an eventual population of some 300 resident animals, two jeeps and diesel fuel, constructed new corrals with thatched shelters and feed and water troughs, and totally restored his dilapidated buildings, putting up an additional building to accommodate volunteers. She also networked for veterinary volunteers and veterinary nurses to come from the U.S., Canada, and Europe, and secured funds to have a much needed well drilled right on the property to provide fresh water year-round. Formerly during the dry season water had to be brought in by cart. Most importantly she established an effective spay/neuter and vaccination program that systematically went to every village and tribal community in and around the Mudumalai Wildlife Preserve, an integral component of the Nilgiris Global Biosphere Reserve. While donations covered her travel and living expenses during her frequent tours of duty that were up to 8-months duration, she never received any salary.

Soon after IPAN was registered in India, Chinny Krishna (with the Blue Cross Animal Welfare Society in Chennai, and the Animal Welfare Board of India, a governmental organization), informed Krantz, who was back in the U.S., that she did not have government approval to receive foreign donations to support her project and advised that she temporarily remove herself, and Dr. Fox, as Managing Trustees until such ('FCRA') approval was granted. He had her wire $5,000 to his personal account in order to put this process on a fast track, otherwise it would take many months. In the interim, Mr. Otter would be named Managing Trustee on a new Deed of Registration that was considered by all involved as a mere formality that would be rectified once FCRA approval was granted, and Krantz and Fox would then be reinstated as per the original Deed of Registration.

When government permission to receive foreign donations was eventually granted, Nigel Otter claimed Krantz was simply an "erstwhile donor" and all that she had built and put into Hill View Farm belonged to him, and that he was the original Founder, and now Managing Trustee and Director as per the new Deed of Registration that he filed with the local government.

Between 2003-4, while breaking protocol and having liaisons with visiting women volunteers (according to the Staff), Otter repeatedly insisted to Krantz that it was not yet safe for her to return to India because she could be arrested for continuing to send funds that were desperately needed for all the animals, without FCRA approval. It was during this time that he was orchestrating the take-over of IPAN. Then in March 2004 he conceived a child with a veterinarian volunteer from Finland. Like all volunteers screened and selected by Krantz prior to coming to work at Hill View Farm Animal Refuge, this woman had agreed to never fraternize with any of the staff, but intimated that her Finnish boyfriend was planning to visit (he cancelled out). Krantz, uninformed by either party of this new development, continued to send funds to support the project until Mr. Otter, after repeated calls to him to send monthly reports on activities and expenditures, eventually sent clearly falsified accounts with a paid-for accountant's signature. He even used Krantz's own personal moneys, $10,000.00, that she wired to him in an emergency after he claimed to have run out of funds to pay staff and purchase animal feed and medicines for an unapproved/uniformed leave of absence, to fly to Finland for the birth of his first child.

Prior to his planned deception to take over IPAN for himself, and the Central government's intervention on the issue of Krantz sending funds from IPAN U.S.A. to support IPAN's work in the Nilgiris, Maneka Gandhi, (then Minister of Animal Welfare and Social Justice), had sought to discredit and defame us calling Hill View Farm Animal Refuge and all programs and projects a sham, amounting to nothing more than 'country retreat for the Foxes, with just a few animals.' The disinformation eventually reached the U.S. and was spread erroneously by a confused Merritt Clifton, through his publication *Animal People*, that we were "getting rich off the backs of India's suffering animals." This was no doubt predicated by the fact that IPAN was perceived as a threat to vested interests in the corrupt Nilgiris area, and to the broader nexus of India's animal welfare community, both governmental and non-governmental. Clifton refused to accept an all-expenses-paid visit offered by Krantz to come and see her animal refuge and learn the truth.

It is now on public record (*The Hindu,* newspaper's "India Beats" magazine supplement, Nov. 19th, 2006, p.7) that Nigel Otter tells the press that he is the Managing Trustee of IPAN, and that 'The animal shelter was originally a cattle farm run by Nigel. In 1999-2000, it took in 60 animals at the request of an American animal lover." Deanna Krantz, the "American animal lover", he never mentions by name, his 'legal' take-over of IPAN as the Managing Trustee being documented by this bogus re-registration that he accomplished in 2004.

On his IPAN web site, he stated boldly: "We wish to acknowledge the initial help given to us by Deanna Krantz and GCC Inc., USA, who unfortunately are no longer associated with IPAN in any way since January 2005." Web sites like Mr. Otter's give the appearance of authenticity, especially when it shows links to other purportedly legitimate organizations such as the World Society for the Protection of Animals, and is filled with poignant pictures of sick and injured animals being treated. The October 2005 'Newsletter' on his web site reveals the kind of emotional blackmail and defamation that we thought wrongly were beneath him where it is written "As most of you know we have been left without any funding whatsoever by our founder and erstwhile principal donor Deanna Krantz since December 2004." He fails to inform that this was a consequence of his own decision to take over IPAN for himself and to blackmail Krantz into continuing to send funds otherwise her animals would suffer the consequences. In fact, she was never the "principal donor", but IPAN's many donors from the US and around the world whom Krantz and I solicited through IPAN USA newsletters and web site.

In continuing to use the name of IPAN, and calling himself the Founder, Director, and Managing Trustee, Nigel Otter has become indistinguishable from those who originally tried to prevent IPAN from becoming established in the Nilgiris, and who sought to get Deanna Krantz deported and her visa revoked. He knew full well that my wife and co-worker, Krantz and I were in the process of selling our house in Washington DC in 2003 to establish a trust fund for IPAN, and that our Wills named IPAN as the primary beneficiary. Our commitment was clear and unequivocal, but he had quite different plans. We can only wish him well, for the sake of the animals, and accept his deception and betrayal, which we erroneously believed to be beneath him, as yet another a lesson in the proclivities of human nature. But he should not be using the name of an organization in existence since 1998, and registered under Global Communications for Conservation Inc., New York, NY, namely India Project for Animals and Nature.

IPAN USA and Deanna Krantz continue to be a presence in the Nilgiris, with the eyes and ears of a network of in-field observers, continuing financial support for Dr. Sugumaran and his staff to provide veterinary services and save the vision and mission of IPAN from becoming yet another tragic and unnecessary extinction due to the deception and betrayal of those who falsely claim to care for animals and the last of the wild. As of January 2015 the beautiful Hill View Farm Animal Refuge was abandoned, save for a caretaker and his dog. Nigel Otter and his veterinarian wife live in town with their two children and are involved in an international program to train veterinarians to spay/neuter and work with the U.K.-based "Mission Rabies" project. Otter was documented giving a litter of 7-day old pups the anti-rabies vaccine in Gudalur on Dec 6th, 2014. No doubt not a single instance, this is against the manufacturer's directions and could potentially undermine the entire anti-rabies project if the proper age-defined vaccination protocol is not followed.

Aside from the sexual proclivities and other 'impacting' indiscretions of foreign volunteers in other cultures and contexts, the slanderous disinformation and ideological sympathy fabricated by those who seek to manipulate and exploit the funding opportunities provided by outside (and generally naïve), philanthropic, humanitarian, animal protection, and wildlife conservation organizations—governmental and non-governmental—is.exemplified.by.Nigel.Otter's.repeated-to-all.statement.that Deanna Krantz "Has left India. Abandoned her animals, and no longer works here." Such emotional blackmail, slander and betrayal, after almost ten years of dedicated work and total commitment to IPAN by Deanna Krantz, should be a warning to all who would invest their lives and hopes in serving the greater good, put their trust in others, and go perhaps where even angels fear to tread: But better to deal with known devils than unknown angels in the knowledge that the road to hell is paved with good intentions!

*The Whistling Hunters Field Studies of the Asiatic Wild Dog, 1984. State Univ. NY Press.

www.ingramcontent.com/pod-product-compliance
Lightning Source LLC
Chambersburg PA
CBHW041454210326
41599CB00005B/245